LEADERSHIP IN A TIME OF CONTINUOUS TECHNOLOGICAL CHANGE

ALIGN, STRENGTHEN, AND MOBILIZE YOUR TEAM

Bar Schwartz

Apress®

Leadership in a Time of Continuous Technological Change

Bar Schwartz
Berlin, Germany

ISBN-13 (pbk): 978-1-4842-6299-3 ISBN-13 (electronic): 978-1-4842-6300-6
https://doi.org/10.1007/978-1-4842-6300-6

Copyright © 2020 by Bar Schwartz

This work is subject to copyright. All rights are reserved by the Publisher, whether the whole or part of the material is concerned, specifically the rights of translation, reprinting, reuse of illustrations, recitation, broadcasting, reproduction on microfilms or in any other physical way, and transmission or information storage and retrieval, electronic adaptation, computer software, or by similar or dissimilar methodology now known or hereafter developed.

Trademarked names, logos, and images may appear in this book. Rather than use a trademark symbol with every occurrence of a trademarked name, logo, or image we use the names, logos, and images only in an editorial fashion and to the benefit of the trademark owner, with no intention of infringement of the trademark.

The use in this publication of trade names, trademarks, service marks, and similar terms, even if they are not identified as such, is not to be taken as an expression of opinion as to whether or not they are subject to proprietary rights.

While the advice and information in this book are believed to be true and accurate at the date of publication, neither the authors nor the editors nor the publisher can accept any legal responsibility for any errors or omissions that may be made. The publisher makes no warranty, express or implied, with respect to the material contained herein.

    Managing Director, Apress Media LLC: Welmoed Spahr
    Acquisitions Editor: Shiva Ramachandran
    Development Editor: Rita Fernando
    Coordinating Editor: Nancy Chen

Cover designed by eStudioCalamar

Distributed to the book trade worldwide by Springer Science+Business Media New York, 1 New York Plaza, New York, NY 100043. Phone 1-800-SPRINGER, fax (201) 348-4505, e-mail orders-ny@springer-sbm.com, or visit www.springeronline.com. Apress Media, LLC is a California LLC and the sole member (owner) is Springer Science + Business Media Finance Inc (SSBM Finance Inc). SSBM Finance Inc is a **Delaware** corporation.

For information on translations, please e-mail booktranslations@springernature.com; for reprint, paperback, or audio rights, please e-mail bookpermissions@springernature.com.

Apress titles may be purchased in bulk for academic, corporate, or promotional use. eBook versions and licenses are also available for most titles. For more information, reference our Print and eBook Bulk Sales web page at www.apress.com/bulk-sales.

Any source code or other supplementary material referenced by the author in this book is available to readers on GitHub via the book's product page, located at www.apress.com/9781484262993. For more detailed information, please visit www.apress.com/source-code.

Printed on acid-free paper

Contents

About the Author v
Acknowledgments vii
Introduction ix

Chapter 1: The ACE Model 1
Chapter 2: Identity 17
Chapter 3: Emancipation 35
Chapter 4: Capability 55
Chapter 5: Autonomy 75
Chapter 6: Leading with Clarity 97
Chapter 7: Leader's Question Guide 109

Index 113

About the Author

Bar Schwartz is an Organizational Excellence and Agile Leadership consultant and coach. Currently, she heads the technology department at medneo, where she supports the leadership in simplifying ways of working, building digital product development capabilities, and enabling further company growth. Furthermore, she works together with AgileLAB to deliver professional, hands-on Agile trainings.

Bar started her career as a software engineer and has over the years worked in leadership roles where she led tech, product, people, and transformations in diverse organizational structures, such as startups, corporations, and consultancies.

With her work, she aims to enable people to create successful change in their work and organizations so that together they can create more meaning, more value, and a better future. Her focus is agile- and technology-dominant environments as they evolve faster than we humans can often adapt to.

Acknowledgments

Foremost, I would like to express my sincere gratitude to Dave Stachowiak and my Coaching for Leaders Academy cohorts for their continuous encouragement and support, which inspired me to write this book. Thank you for teaching me that every book starts by writing one line.

I would like to thank all the great leaders I met along the way, from whom I learned so much. They inspired the case studies in this book. Thank you for allowing me to support your journey.

My sincere thanks also go to Minna Paananen, Vera Hillmann, Karo-Lyne David, Iris Henne, Boris Schulz, Anita Ripke, Vanessa Englert, Filip Moriau, Philipp Zupke, and so many other great friends and colleagues for supporting my journey. Thank you for every thought-provoking conversation and feedback.

Finally, I would like to thank my family, especially my mother, Dorly Schwartz, and my father, Pini Amar, for raising me to be the person I am today.

Introduction

To become a leader in an organization, it takes a lot of ambition, competence, and discipline. Leaders set the pace and drive results. As such, they need to be the best in what they do and set the standards, lead by example, focus on impact, and get things done. If they do so, people will be motivated, empowered, and engaged. This will drive better economic success, which will lead the organization to realize its vision. They will win the game!

Wait, what game?

My name is Bar Schwartz and I used to believe that I needed to be all those things to be a leader. I needed to be an expert, the best in what I did. I needed to have the answers to all questions. I needed to overcome myself and my emotions so I could be resilient and disciplined. I thought that it was all about me becoming the best version of myself because only then I could lead. I believed that no one wants to follow people who are not good enough, so I better be.

To be precise, I used to believe all of that, and then I realized it didn't matter. How does becoming an expert in my field help the people I lead? It doesn't. Don't get me wrong, I am not saying having expertise is not important. I am also not saying that leaders should give up their learning and development journey. I am a true believer that we ever evolve and grow. What I am saying is that you being an expert is not going to help other people to develop their expertise. Yes. Even when in my head, I wanted to help, be resourceful and needed. In reality, I covered my inner insecurity by telling myself that I am ambitious. I covered my perfectionism by telling myself that I am building my competence so I can help more people with my wisdom. If I have all the answers, people will want my help. If I work hard enough, if I am disciplined enough, my work will speak for me so I don't have to speak for myself. I am an expert. Experts are resourceful. People follow experts. Don't they?

For years, I evaluated myself by my achievements. I needed to be the best. When my work was celebrated, I was celebrated. When my work was criticized, I was criticized. I cared very much about my competence. It was not good enough to be good enough. However, how can you be the best without everyone else being worse? How can you allow people to explore, evolve, learn from their mistakes if they learn from you that nothing is good enough? I set my bar so high that even I couldn't reach it. By doing that, I made sure everyone around me felt that they were also not good enough. One can't hold themselves to different standards than they hold others to.

Introduction

I didn't come up with this belief on my own. I am the type of person who listens to others more than they listen to themselves. Hence, I learned it from the role models I had. People like my mother, teachers, and most of the mangers I had since I started my career back in 2009. They validated me for being smart and competent. It was the most important thing about me. Maybe it is how my brain is wired or maybe it is the trap many of us fall into when we care about what people think. I don't know. What I do know is that it was not only my belief that if I am the best at what I do, and people know that I am the best at what I do, I will be rewarded for my expertise and people would want to be led by me.

So, I was the best and I was promoted. I had no trouble finding new jobs, better roles, and generally selling my expertise to different employers. I became a team lead in 2014 because of my ambition and expertise. It was less than five years into my career and only one year after I officially graduated with my bachelor's degree in software engineering. I was so proud to be acknowledged for my hard work. So imagine my surprise when I discovered four months into the job that I was not naturally a great team lead, that I couldn't handle people. No matter how great I was as an engineer, being "the expert" didn't make me a great leader.

Every new leader wants to demonstrate that they can bring so much more to the table. My ambition led me to attempt to do the same. From the first day of taking over the leadership role, I started changing things to fit my ideas and style, "hoping" that the others would follow suit. I did it because I thought I knew the work best. After all, I was hired for my expertise. All those changes, I thought, I did for them. I cared; I wanted to help. Also, when I obsessively identified everything that was done wrong, many things improved in a relatively short time, which made me feel great about myself. I was sure people were happier. However, how did my team feel? In short, they felt abandoned.

As inspiring as this idea of strong leadership and devoted followers is, it is a flawed system that has led to long-term setbacks with initiated projects, including losing the people the leader led on the way. After all, why would people follow you if it is all about you and nothing about them? Even if things improve, what makes you believe that things improved for them?

Thinking about people as followers is not just flawed, it is dangerous. Why? Because it works. You got your leadership role because you are smart. You are competent. You get things done. You can create short-term results. However, in the long-term, this mindset slows you down, disconnects you from the people you lead, and limits your scalability and adaptability capacity. Why? Because it assumes that everyone else on your team will do as you tell them to. It assumes people are motivated in the same way as you. It assumes people think, feel, and understand the world in the same way you do. It assumes that if you set ambitious goals and control the process, people will follow it willingly, no questions asked. After all, you are the expert and they all want to learn from you, right?

Introduction

Let's take a step back and talk about leadership. What is leadership? It is the ability of an individual or an organization to guide a group of people towards achieving a certain outcome. In the business world, we often confuse leadership and management, the process of dealing or controlling people and things to achieve a certain outcome. We even tend to call managers, leaders. For example, my first team manager title was Quality Assurance Lead. I had management responsibilities because people reported to me. Nevertheless, leadership and management do not have to go together. One can lead without having management authority, and one can manage without leading by merely demanding. However, they both aim for an outcome. We, you and me, are experts in those outcomes. This is why we got the job.

When we speak about outcomes, most organizations speak about business results. This is often measured by the profit the business makes because if your business is not a non-profit organization, you have to show that your revenue is higher than your operational costs to remain in business. Therefore, regardless of whether your organization is serving a cause, selling a product, or providing a service, the underline trends every organization aspires to are to increase revenue, reduce operational cost, and mitigate risks that could disrupt these goals. As a manager in an organization, you are given the authority to achieve these goals by ensuring that your people contribute their part. As a leader, you are the custodian of the vision and your responsibility is to enable people to realize it—not for you, for themselves.

Leaders lead people, and people create results. Leaders mobilize and inspire team members to channel their energies towards the actualization of a vision or a goal, but people are *different*. Everyone you work with has a different personality, feelings, thoughts, skillset, and more importantly, they come from a different work culture, even if they worked in the same country, city, or company. They see the world through their unique lens. They move to action differently, get inspired differently, and manage their energy differently, meaning that what works for you might not work for them. You are an expert in how you get the best results, not how everyone else can.

If you dismissed the previous paragraph by thinking something along the line of "my organization is mostly homogeneous, so I do not have to worry about that," well, maybe. But maybe not. Will it always be like that? I am not sure. As location boundaries disappear, globalization brings a whole new challenge. These days, people can relocate easily or work remotely with an international team. It means that organizations can hire the best talents regardless of where they are located! It means that diversity is the reality of many organizations, not an ideal or a desire. Thinking globally and appreciating and seeking diversity of background, thoughts, and expertise are going to become the new normal. Like it or not, if you want to lead, you will have to lead people who are different than you.

Introduction

So is this whole book about telling you to not be an expert? Or is it about how to get results when you lead people who are different than you? Maybe it is about how to stop thinking that you know best and let people come up with their unique way to solve problems. Not exactly. There are so many books about that. Yes, this book is going to help you with that in a very practical way. However, this book is also about you becoming a different type of leader. It is going to challenge you to let go of the strong expert that you became because being a great leader isn't just about achieving a set of goals. It is also not about being the smartest person who knows how to get things done. While these traits are important to you, being a leader is more than that. Being a leader is about bringing your people along on the same path and enabling them to grow to deliver these goals, even if you are not there anymore. You are not a great leader if the organization is successful while you are there; you are a great leader if it remains successful even when you are not there.

If you are like me, your first thought might be that your results will speak for themselves. Your results will last and represent your mark on the organization. Maybe. I am not sure if anyone remembers the results I created in each organization where I worked. Moreover, I am confident some of my work had to get redone because things changed. For example, I worked as a Product Manager at an organization called HERE Maps. When I started working there, they were owned by Nokia. Less than a year after I joined, my whole team was transformed because Nokia sold all of their Windows Phone devices to Microsoft. Then, less than two years later, they sold the whole company to a consortium of German automotive companies. At some point, I stopped counting the number of organizational changes we had. Nevertheless, each of those changes transformed everything, including organizational structure, processes, roles, strategic focus, and culture. What remained was not the work I did, it was the people.

Continuous change and uncertainty are now the reality of many other organizations. It doesn't have to be as extreme as HERE Maps. It can also be a result of normal technology evolvement or changes in the market. It is a common belief that only technology companies have to adapt quickly, that only software teams desire to become more agile and accelerate their product or service delivery to remain relevant in today's fast-moving world. Other sectors deal with the same challenges. This can be due to digital automation of roles and processes that used to be manual, or an unexpected circumstance such as the COVID-19 pandemic, which disrupted the way people live their lives. Business models are reinvented every day, technology solutions become commodities as technology evolves, and so many large corporates are slowly going out of business due to creative companies, mostly startups, that disrupt the status quo in ways larger corporates are often too slow to adopt. You might hear about "Agile" or "Lean." Most organizations realize that having an organizational transformation is the new status quo. In such a reality, results are important, but they don't last. People do.

Introduction

Change brings a new layer of complexity to leadership. I acknowledge that I raised the bar very high. I am telling you that you can be a great leader without being the expert in what you do, without telling people how to do their job, and even without having the ambition to create amazing results yourself. I am telling you that you can work with people who are different than you. You can ensure that everyone is onboard with your vision and able to realize it by themselves for themselves. Lastly, I am telling you that you can achieve all of this while your organization is continuously changing and you can ensure that lasts even if you are not there anymore.

Can you imagine a world where the people you lead see themselves in that vision you are trying to achieve? Where they feel personally accountable for it, accepting it as a desired reality for themselves? Where they are so aligned that you don't have to tell them what to do, they just do it? Where they unblock themselves rather than wait for you? Where you don't feel the pressure of always having to be the best? Then, this book is for you. If you can imagine it, you can build it and you can be it. It worked for me, but I am no expert.

This book is both inspirational and practical. I wrote it to inspire managers and leaders of cooperations, businesses, and organizations (large, medium, and small) to operate as an inclusive system with the people they work with. As a businessperson, you know that everything achievable in business comes with a process that leads you to results. This process was developed based on my experience and it helped me to transform the way I lead. Those were my results and the results of other leaders who followed the same approach. Our results are described in this book.

I share these results with you, hoping that they will provide a practical framework to transform yourself while you align with the people you lead. As you read, you will discover that the framework focuses on generating clarity for yourself and the people you lead. Clarity is the baseline for alignment.

Furthermore, you will discover approaches on to how to connect with your people, as well as how to become an inclusive hands-on leader who leads an organization that works together collaboratively, even when you are not there.

Enjoy your journey.

CHAPTER 1

The ACE Model

Leadership That Brings People Along to Clarity and Alignment

Digital technology, digitalization of traditional business models, and technology-led companies have affected the way everything is done and how we perceive certain ideals, especially in the world of business. It does not matter if you are working in an organization that develops software products or not, digital technology impacts you. It impacts your customers. Therefore it impacts you. It impacts the expectations of your stakeholders, namely customers and employees. This is a digital age. The impact of digital technology is going to grow and accelerate every year. Especially for the young, utilizing digital technology is not optional, it is a commodity. For your business, being adaptive is the new status quo, not a trend.

Anyone who wants to successfully lead individuals and teams in this digital age needs to use a potent combination of entrepreneurial mindset and a diversity of perspectives to problem-solve effectively. Some will call it "the future of work;" others will include it under "agile" or "lean" transformations. Fundamentally, it is about change and how effective you are in leading people through change while utilizing your existing resources.

© Bar Schwartz 2020
B. Schwartz, *Leadership in a Time of Continuous Technological Change*,
https://doi.org/10.1007/978-1-4842-6300-6_1

Why Are You Here?

You are, most likely, a leader in an organization that is going through a change or about to go through a change, probably due to digital, agile, or lean transformation. You see this friction between the world you know and the world you are getting to know. Processes are not enough. People keep pulling things in different directions. They are not working on the same things or towards the same outcome. You care about it. Maybe it even frustrates you. Predominantly, you want to solve it.

Me too! I faced this problem so many times in my career. As a change catalyst, I often joined organizations in transition. Whether I was the change leader, coach, or enabler, it was always the same problem: lack of alignment due to lack of clarity on roles, the work people are expected to do, the outcome of the change, the product vision, the company strategy, the company values, and more. So, what do you do? How do you solve this? This book is about alignment. It is about how to create it and how to get your people to contribute best during a time of change.

To achieve this, I developed the ACE model. It stands for

- **Emancipation**: Set people free from having to constantly rely on you.
- **Capability**: Enable people to get the job done.
- **Autonomy**: Enable people to own their work.

In short, it is a model that aspires to support you in clarifying what is it that you want and align better with your people so that they can work towards that goal. Ideally, I hope you have a journey while reading this book. I hope it will help you to understand where you are and how to get what you want.

The ACE Model

The ACE model is an exploration and clarification process that you, as a leader, can go through with the people you lead when you kick-start a change process. It is tailored for digital change processes such as agile and lean business transformation. The goal is that everyone is clear and aligned on what is needed in order for a transformation and change process to succeed. The model exists to support you in leading this process. Furthermore, as you clarify, you will discover the prerequisites for your change.

This model is different from other models out there because it focuses first on how you need to think. It is not a recipe book with step-by-step instructions. However, it is an actionable framework that offers you a perspective and concrete steps you can take based on your unique situation.

The ACE model process focuses on achieving *clarity*. It consists of steps that build upon each other and it poses questions that will help you lead with clarity as you take everyone along with you towards alignment. In many ways, it is a coaching process because each question is meant to start a conversation that creates awareness for all individuals involved and generates learnings that will enable you to succeed moving forward.

Forward is subjective, though. You will have clarity on where you are and where you can go rather than a concrete plan on how to get results. It means that you will have an alignment, not necessarily an agreement. There might be a huge gap between your desired outcome and where you are. Nevertheless, if you do not have clarity, you do not have alignment. It is a matter of luck if you get there. Hence, moving forward with clarity.

Clarity As the Core of the ACE Model

William Arthur Ward once said, "If you can imagine it, you can achieve it. If you can dream it, you can become it." Clarity, in the context of this book, is all about translating an abstraction into something precise enough to imagine so that you can take the right actions to become who you need to be to succeed.

Change occurs in your life, either intentionally by you or externally by your environment or others. We are all familiar with life changes such as starting a new job, dating a new person, and moving to a new place. At first, we might have an idea of what is going to change. However, clarity, what changed and what is expected of us due to this change, evolves over time.

When you start a new job, you ask questions such as "what is my role?" and "what does good performance in this organization look like?" When you start a new relationship, you ask questions such as "what do they like or dislike?" and "how do I fit with this new other person?" When you move to a new place, you ask questions such as "what can I bring from my previous place and what should I buy new?" and "where should I shop, work out, park my car?" What questions do you ask when you kick-start a digital change process?

It is normal for us to ask questions to gain clarity about a job, relationship, or a move. When we have sufficient information, we feel prepared for the change. Then, when we go through the change, we feel more confident, aligned, and enabled. We know what to do.

This is not the feeling most people have when an organization, department, or team goes through a digital transformation. The more common feelings are fear, anxiety, and confusion. Mostly, what needs to change comes down to how we need to change and why we need to change. Furthermore, most companies have a document or training that describes the change, primarily focusing on the new job titles, work processes, and tools. However, these documents often remain either too abstract or too focused on process-related details. Therefore, you might understand the concept and the overall process but have no idea how your day-to-day work is going to change and what is expected of each individual to make this change successful. For example, you might want people to work together collaboratively as a team on a concrete product. Also, you might want individuals to help each other instead of starting new work in their function. However, due to traditional understanding of roles, performance, and deliverables, people might think they did a good job if they finished their part of the work rather than ensuring the whole team finished a concrete work increment and created value.

Clarity is a state, not a destination. Therefore, to achieve it, you must ask the right questions so that you understand your situation and yourself better. You also must keep asking questions as your situation changes. It's the same for the people you lead: clarity is about having a conversation about what matters. This is the core of the ACE model.

Identity As Step Zero

We all see the world through our own lens. Thus, it is common that the same situation is perceived differently by different people. For example, one person may be excited when it starts snowing while another may complain about the cold temperature or the mud. It depends on our attitude towards snow.

Dr. Wayne Dyer once said, "If you change the way you look at things, the things you look at change." If you learn to see a snow storm positively, you may welcome it. The same happens with change. When you see it positively, you are likely to welcome it. However, if you do not, you are likely to resist it.

Knowing yourself is the foundation to leading change successfully because we tend to support changes that are aligned with what we perceive positively. These are changes that we are willingly investing time to learn and adapt ourselves to. So, if you hate snow, how likely is it that you will support relocating your team to a city where it snows most of the time? Also, what will be your attitude towards the people who love snow when you hate it so much? Something to think about.

Identity is also a state, not a destination. You are not the same person today as you were ten years ago. As you grow older, you naturally engage in a process of getting to know yourself and start reflecting on who you are. That enables you to both support changes that are aligned with your core values and understand the differences between others and you. Hence, identity is step zero, not step one. You can theoretically lead a change without knowing yourself, but it will be harder to lead yourself and others towards a future you do not really want.

Starting with the End in Mind

The model is called the ACE model, but the order of the book is Emancipation, Capability, and Autonomy. Why not ECA then? It's so that you always keep the end in mind. The final state is for individuals to be able to work autonomously either by themselves or with their team. To achieve this, they need to be capable of doing so. Also, they need to be emancipated. I will explain each of these concepts in their respective chapters.

Before you dive right into the model, you must first understand the context you operate in. Hence, what changed in digital technology-led environments?

The Impact of the Digital Age on Our Work Environment

The use of digital technology has led to more businesses trying to reinvent themselves with novel digital services, products, and ideas that suit the expectations of their teeming customers. Many new digital businesses emerge from looking into a problem and realizing that there is a better way or a different way to solve it. The focus is on customer experience. Where there is a bad customer experience, there is an opportunity to innovate with a new product or service. People pay money for products and services that improve their lives. For example, we once had to make a trip down to our local bank and speak to a teller to check our account balance. That was very time-consuming for the customer. It was also not a very effective use of the teller's time. When call centers were introduced, we could also call our banks, which saved time and reduced the number of customers in the bank every day. Now, we can check our account balance via the bank website or mobile app 24/7 without interacting with a teller. Banks that were slow to offer an online platform lost customers rapidly.

When customer experience improves, the expectation of what is possible also grows. It is very similar to when you get a raise. At first, you are excited. You think you will be able to save more money, pay your debts, and finally get the new phone you wanted. A few months later, you realize that even with the raise, you save about the same amount of money. Now you want another raise. Why is that? You adapt your lifestyle to the new income. What you once considered new and exciting becomes the new normal. Therefore, the more normal the new improvements get, the less meaningful each one becomes. When the first iPhone came out, it took the world by storm. It was exciting and everyone wanted one. Now, we have come to expect a new iPhone model twice a year, every year. The releases have become unremarkable. It takes so much more to delight customers. It takes much more to be considered the best.

The desire to become a leading business that is recognized for excellence has encouraged business leaders to increase their speed of delivery. This is a result of many companies believing in the first mover advantage. Hence, there is no time to waste because customers are not waiting. It is easier than ever to switch providers in almost any type of industry. If your business is too slow to adapt or is unable to provide constant innovation and better customer experience, you are out of the market.

There is some truth to this. Speed is important. I hear more and more about companies that have added "move fast" to their company values. "No time to waste," senior leaders say. "The competition is not waiting." They need to move faster, adapt faster. This has become a necessity. However, do you know what you are trying to achieve? Does your team know?

Businesses nowadays operate under so much uncertainty with so little time to think things through, plan ahead, and ensure everyone is involved or onboard. Then, in practice, they waste so much time trying to move fast that they rarely realize that they have no idea what they are trying to achieve. That people are not aligned and their results are not good enough. So, what do you do?

It depends. Many companies take a journey to digitize themselves, automate their manual data-gathering processes, or adopt agile or lean strategies to accelerate and iron out delivery processes. This, in turn, enables the organization to shift priorities when new information arises. However, it is not a magic solution. Also, depending on the organization starting point, and leadership capabilities, it can take a long time, fail, or not happen at all.

Note "Agile" or "lean" are both methodologies that enable organizations to work in a way that maximizes the value created to their customers. Agile focuses on responding to change while integrating feedback that ensures the work is valuable to the customer. Lean focuses on minimizing waste by continuously optimizing the process and focusing the work on what creates value to the customer. The right approach for your organization depends on your unique business context, market, and organizational culture.

Any organizational change is a transformative process, large or small. If people need to change the way they work, they often need to change the way they behave, think, and feel. Depending on the gravity of the change, it can be easy or hard. Do you remember the last time you changed a habit? It can be anything from waking up earlier, going to the gym, or eating more vegetables. What did you do? Did you force yourself out of bed? Did you ask your partner to remind you? How long did your change last?

Changing a simple process in an organization is the same as changing a habit. Let's say that a company has encouraged everyone to use instant messaging software instead of the phone. For those who like to talk on the phone, it may take them a little while to break the habit. For others who do not like to talk on the phone, the change will be easier. It's the same as with any other habit. You can achieve the behavioral change by force or by enablement, meaning you can tell people what to do and put systems, processes, and rules in place to force them to change. Or, instead, you can communicate an expectation and work together with people to design the systems, processes, and rules to enable them to change.

There will always be people that do not see the value in the change. For them, none of this will work. You will have to support them in finding this value for themselves. For example, I might not enjoy waking up early for the sake of waking up early because I am not an early riser. However, understanding that it will give me extra hours to drink my coffee and plan my day will get me out of bed. Nonetheless, it won't get everyone out of bed. Each person must find what works for them.

The transformative process in organizations of this digital age entails the adoption of new values, cultural ideas, and different means of working. Often, they are not in alignment with the current way things are done in the organization. Otherwise, it would not be a transformation. The most common complaint I hear from people during a transformation is that they always do things in a certain way and it works, so why should they change? Well, it may work for them, but does it work for the organization, the product, and the customer? If it does, don't change it.

Also, you can only drive one change and have to be consistent about it. You cannot do things the old way and also the new way. Thus, if you are a leader (or an aspiring leader) in an organization that is going through a transformation, you must present a united front with your fellow leaders and drive the same changes. If you do not agree with the vision, then you face the difficult decision of either supporting the transformation anyway or kick-starting the hard process of clarifying what you want versus what is possible. You must align. Sometimes you will find the middle ground and sometimes you will decide to seek employment elsewhere. Yes, this is harsh. This was my main struggle in many changes I led because I am not the type of person to support something I do not fully believe in.

Transformations can be driven top-down or bottom-up. In simple words, if your CEO drives it, it is top-down. If the people drive it without management support, it is bottom-up. Successful transformations require both top-down and bottom-up support to align. It does not matter who initiates it. What matters is ensuring that everyone is working towards the same future. This is crucial. Very few one-sided transformations succeed, again because you cannot drive two competing changes at the same time.

The main reason for this is people. People make or break transformations. In every transformation, most people go through a harrowing journey that leaves them unclear, overwhelmed, and confused, mostly because they keep receiving inaccurate, competing, or unclear information about what is changing, why, and how. Almost every transformation raises questions around roles, responsibilities, and information transparency. It is scary to not know what you do not know. For most leaders, this is especially stressful because all of these questions come to you. If you also have the expert mindset, where you believe you should have all the answers and should never admit that you do not know something, imagine the stress of not having these answers or, worse, having the wrong answers.

In order to lead in the digital age, you must acknowledge the constant uncertainty you operate in and strive to continuously seek clarity on all sides so you can support the people that look up to you for answers.

The Impact of Lacking Clarity

Every leader who is confused seeks only one thing: clarity. I was one of those confused leaders. My background is in software engineering and I have worked in software companies my entire career. So, when I joined a consulting firm in 2017, it was a whole new world for me and, apparently, for them as well.

I joined one of the top tier consulting firms as an Agile Coach, a digital expert. It was considered a new role, a new function, and a new team setup. Digital was new to them. Moreover, working with digital experts was new to them. This lack of familiarity led to a misaligned perception of what the new digital roles were about.

Digital experts were hired to support clients who were reinventing their core utilizing digital capabilities. Hence, digital engagements were staffed with a mixed team of strategy consultants and digital experts to ensure they brought the strategic value consulting firms are known for, together with the training, coaching, and implementation power of the newly joined experts. Ideally, with all that expertise onboard, we could help the client to kick-start their digital journey.

As an Agile Coach and a digital expert, my engagements included joining the initial stages of setting up the first digital team of the client or conducting introductory workshops to align everyone on the new culture, principles, and ways of working necessary to operate in a digital working environment.

I thought I was there to both help them to implement a digital product and coach the client. In practice, we only focused on the product implementation and generating documentation.

What happened? The firm had best practices and blueprints on how to do everything. This is the typical way many consulting firms work. The deliverable of a consulting firm is often an elaborate slide deck describing the problem and a step-by-step solution. In a typical consulting engagement, this works well. It does not work for a digital engagement.

Here I was, staffed on a client engagement, seeking ways to clarify what the client needed and coach them through this transition. At the same time, the engagement manager kept asking me to prepare documents explaining everything we were doing. Before every client workshop, I was so busy preparing elaborated slides based on existing blueprints rather than working together with the client to define what they wanted and needed.

At my first engagement, I played along because I was new, and I wanted to learn. During the second one, I was lucky to have a collaborative partner who understood that I was not familiar with the role. The third engagement, however, was not as great. The partner, engagement manager, and I were in constant conflict. When the same situation happened again in my fourth engagement, it was also my last. Could it have gone differently if we were aligned?

We See a Different World

The firm was going through a transition, learning to accommodate both the traditional consulting world and the new digital world. As mentioned, new roles were introduced. Therefore, people of different profiles were hired. People like me, who worked for digital companies previously, mostly software development companies, were added to the mix.

There are fundamental differences between working for a typical software development company and a consulting firm. It starts with the type of clients they work with, how they communicate, and what is important to them during an engagement.

The most typical client of a large top-tier consulting firm is an enterprise. These types of large corporations tend to follow a more traditional hierarchy, long planning cycles, and top-down communication style primarily, if they are not familiar with digital. Therefore, consulting engagements are well-scoped with clear milestones and a fixed timeline. Software companies, however, tend to have a very chaotic atmosphere, a changing scope and an estimated timeline.

Furthermore, work and information are distributed differently. As a contributor in a product development team at a software company, the work is codependent. When done right, we work as a team. Hence, everyone is accountable for the outcome and everyone is working towards it. In consulting engagements, however, there are multiple work streams. While they can be dependent, they do not have to be. Each role cares for a subset of the work. Divide and conquer. The engagement manager oversees it all.

Output is important. If you have nothing to show, you did nothing. In software companies, the product is the output. In consulting engagements, slides, documents, and data models are the output.

When I was staffed on a digital engagement, it typically followed a common blueprint. Those blueprints were new to everyone involved. Also, as mentioned, the roles were new so not many consultants understood my role then. In their defense, Agile Coaching and coaching in general are not regulated professions. Also, I failed to clarify it. However, the biggest challenge was that everyone assumed they understood it, which made it harder to identify the understanding gap and make any judgement on what my role really was.

After my first engagement, it became clearer and clearer that our hybrid team had to create a bridge between two very different mindsets while engaging to shift the client's mindset. We had to learn how to work together as a team rather than each person driving their own output.

I had to learn to explain what I did, what I needed in order to do my job, and come up with a quantitative way to measure what I did, even if the real impact of the engagement would arrive after the whole team was gone.

Yet, I was not as aware of all of this then as I am now. Clarity was lacking during most of my early engagements. We managed to overcome it through painful conflicts.

We argued on silly things such as the number and quality of slides we created. Often, we argued on who should be involved in what stage of the engagement to ensure the right scoping, resources, and timeline. It was not the best utilization of our time, energy, and capabilities.

Taking Responsibility for Clarity

"It was not about me. It was not about them." People tend to attribute personal traits to others when they are under stress. You can call it a victim mindset or a blaming culture.

All I know is that I was under so much stress. The need to perform in this consulting world led to my desperate attempt to adapt myself to this perceived consulting world without understanding what I was adapting to. I thought I needed to have all the answers, which stopped me from welcoming feedback or support. Surprisingly, most people around me shared this problem, which did not make it easier to address.

One engagement led to the evolvement of the ACE model in my mind. As a matter of fact, it was one specific incident where I messed up in front of the client. I did not understand what they needed.

In the meeting with the client, we invited the key internal stakeholders in the organization in order to introduce them to the concept of agility and to kick off the engagement, ensuring everyone on board understood their role in it.

Long story short, the outcome was very different from what we had in mind. First, the client perceived the workshop to be too general. Second, the activities I chose, including games, were perceived as childish. Third, we had too many slides, so a lot of the content was irrelevant.

Surprisingly, I tried to do it right. I worked hard to prepare slides that were in alignment with the blueprint. My inner voice, however, nudged me that doing those slides was a waste of time because "agile" workshops tend to be collaborative. The slides' purpose, in this context, was to keep track of the agenda. Thus, my deck was not very thorough.

The week before the workshop, I presented the first version of the workshop to the partner. There was no information about the engagement itself, the objectives, the new roles, the new process, the deadlines, or how to provide feedback. I was not familiar with those type of kick-offs. Therefore, when the partner reacted with "I expected more," I took it personally as a judgement on my capability to prepare slides for a workshop. I was not aware I was not as thorough as I should have been.

Since I took things personally, the engagement manager ended up filling the gaps of my slides deck. That, due to the deadline, was done without aligning with the client. Finally, there we were, completely misaligned. I felt stubborn and undervalued.

The truth is that it could have been avoided. It was too easy for me to attribute the lack of clarity to the people than to evaluate my environment, my role, and my tasks. Instead of getting so demotivated by the gap between what I thought my role should have been and how it was in that environment, I should have clarified. Also, instead of trying to behave as I thought I should have, I should have clarified. We could align.

When do you find yourself blaming people rather than properly evaluating the situation and the problem? For me, it was a conscious choice to stop feeling like a victim and properly assess the situation. The ACE model emerged as a result of this.

Shifting the Mindset

It took me months of coaching to shift away from feeling like I was the victim. It was hard to let go. It was only when I eventually took responsibility for my own clarity that I was able to let go. When I understood that I am the only one who knows what is unclear to me. I also accepted that convincing and forcing others to accept a different mindset does not work. I repeat: It does not work. From their side, everything was extremely clear and we should have been aligned. It did not matter if I was right or wrong. Most of my team worked for that consulting firm for years. They knew exactly what to expect of an engagement kick-off.

So, what should I have done? The solution was neither being stubborn about my way nor adapting myself completely to the "traditional" consulting way. It was somewhere in the middle where I understood them and they understood me. Together, we could define what collaboration between two mindsets could and should look like.

As the expert in my domain and the driver of this workshop, it was on me to lead it and lead them. I was responsible for filling in the details and clarifying together with them the following questions so that we were all clear and aligned:

- What is the purpose of such a workshop?
- What is the culture of the client?
- What is the most important thing right now to get done with this workshop?
- What is my role in this engagement?
- What is the knowledge we need and who has what knowledge?
- What do I bring to the table?
- What do they bring to the table?
- What is in my control?
- What would give them the confidence in me?
- What expectations do they have of me?

Gaining that level of clarity would have given me the ability to judge what was possible and what was not during this engagement. Instead of getting frustrated and taking things way too personally, I could better evaluate the fit between my identity and this type of role in that type of an environment. I could emancipate myself through clarifying the context we operate in, and by answering the questions such as, What are we here to do? In what way? Also, I could assess my capability to do this role and evaluate better what I needed to learn and why. Lastly, I would gain the trust of the people in charge to work more autonomously. I had to change my way of thinking.

Eventually, I took responsibility for my leadership role in that engagement. Only then did I realize how many times I had been frustrated by the way leaders communicated, especially during a time of change. As an engineer,

I often worked in organizations that were not clear on what they wanted to achieve and what they expected their people to do. The size of the organization or the type of project, product, or service did not matter.

I also had to assume good intentions. I realized how all of the leaders I worked with thought they were extremely clear. It was never their intention to confuse me. Often, they struggled to understand why I was still so confused. It was uncomfortable to realize and admit that as a leader myself, I was not clear. Everything was clear in my head, but no one could read my mind. It is called the *expert trap*, where experts provide advice, direction and solutions without fully understanding the context, level of understanding, and expectations of the other person, assuming that because they know what they are talking about, everyone else should know too.

As an expert, the only clarity you will often care about is your own. You do the work, so you want to ensure you understand. As an expert consultant, you are still responsible for your own understanding because you do the work for someone else. As a leader, you are responsible for everyone else's understanding because they do the work. Moreover, if their work is interconnected, you are also responsible to align it all together.

Alignment is the outcome of being clear on what is expected of you and how your work contributes to an outcome. Clarity is the outcome of the effort you put into a process of clarifying through questioning; there are no shortcuts or magic formulas. It is all about asking questions and seeking answers. It takes time. Some questions feel extremely obvious. However, once you ask them and hear the answers, you understand that even the most obvious things to you are not always as obvious to others.

Think about it as if you are playing a team sport. You have a team of players with one shared goal—to win the game. Each player is different. Depending on the game, they will have different roles. For example, European football has four key roles: goalkeeper, defender, midfielder, and forward. Each role requires a different focus and skillset. During a game, each player keeps their position and aims to be the best player they can be in their role. At the same time, the goal remains clear and aligned by all: to win the game. Every team has a coach. The coach used to be a player and they share the same goal of winning the game. However, when was the last time you saw a coach jump onto the field in the middle of the game to block a goal? Never. Their role is to support each player to become the best player they can be, to support each player to see how their actions influence other players' actions. They ensure that all players are aligned on how to win the game because there is no value in being the best player on a losing team. Similarly, there is no value in being a lead of an underperforming team of experts. If every player plays for themselves, no one plays to win.

Not every team member plays to win. It is common in professional environments for people to have different motives such as fun, money, and learning. According to Gallup's World Poll, only 15% of the world's one billion full-time employees are engaged in their job. That leaves 85% of them unengaged and potentially hating their job, and often, their boss. This is a very concerning statistic given that we invest so much of our life in our work. Also, since you are reading this book, I assume you want to be a better leader and you care about your work. Hence, if you want to have a team that plays to win, it is on you to create the environment where people want to play and win. As in team sports, if everyone in your team is doing their own thing, regardless of the game or each other, would you like to play? Would you like to win?

Recap: The ACE Model

Change takes time. It took me over a year to conceptualize the ACE model. Then I had to test each of the concepts to ensure that they were practical and relevant. I refined the model further while writing this book. I am sure it will forever evolve as I grow and mature in my life and my career.

The ACE model, as explained earlier, is an exploration and clarification coaching process to create clarity for individuals and teams during uncertain times, particularly times of change.

As you can see in Figure 1-1, the ACE model starts with identity (you, and your utilization of the model) and ends with the goal of achieving clarity. However, as explained before, clarity is a state, not a destination.

Leadership in a Time of Continuous Technological Change

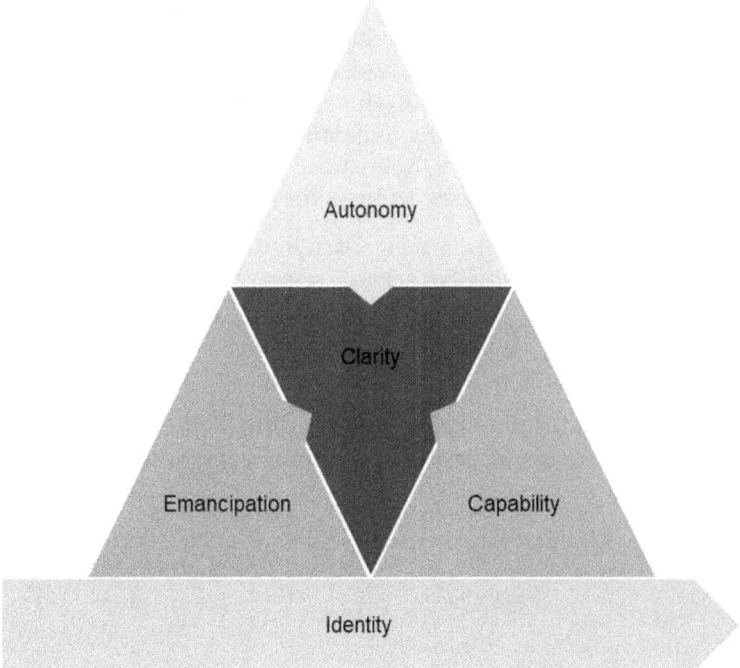

Figure 1-1. The ACE model

Between the beginning and end, you go through the steps of examining the parts of ACE:

- Emancipation
- Capability
- Autonomy

The process is interactive. Each step brings people together to answer questions that help to ensure everyone is on the same page, working toward the same outcome, and realizing the same vision. The ACE model is meant to support you in finding clarity, at the moment for a concrete project or task. Moreover, it is also meant to support you to maintain clarity as information changes or evolves. It is not a model you go through once; it is a compass to check every time things become unclear to you or anyone you lead.

Chapter 1 | The ACE Model

Depending on who is seeking the clarity, it is recommended to go through the model with another person. For yourself, that person can be a coach or your direct manager. For the people you lead, involve key stakeholders, regardless of if they report to you. If you use it with a team, ensure that everyone in the team speaks up. Feeling heard is the foundation here. Hence, if you speak 20% of the time, you are doing it right. Remember, your role here is to understand the other person before you seek to be understood.

Now that I've gone over why having clarity and alignment is so crucial to success, let's look at how you can achieve it with the ACE model.

So let's commence the process starting with *identity*.

Please note that Chapter 7 contains a Leader's Question Guide. This guide contains sets of questions that help you get the best out of the ACE model after reading the book.

CHAPTER 2

Identity

Who Are You in Your Context?

Identity stands for the way you see yourself. When you are asked who you are, your identity is the labels you associate yourself with. These labels typically start with "I." For example, I am an Israeli, I am a woman, I am an empathic person, I care about people and their well-being, and I am results-oriented.

Identity is formed throughout our lifetime. It starts early on when we are infants and matures during our adolescence and early adulthood. Some people experience their identity to be very adaptable. When they look back, they do not recognize the person they used to be. Others may experience their identity to be stable. There is no right or wrong. You are who you are.

Our identity is influenced by our context. Ultimately, we are social beings so the groups we associate ourselves with have a strong impact on how we see ourselves. When our identity is accepted positively by our environment, we are in harmony. So, if you are a logical person surrounded by other logical people who value logical thinking, you are in harmony. You feel accepted. You feel valued. It helps your self-esteem. However, when we are in environments where our identity is not aligned with the people around us and is negatively perceived, we feel misunderstood. We may even go to battle to convince the world that we are right to be who we are. Conversely, we may feel like we have to conform. Depending on how stable or movable your identity is, conforming can be an easy or hard thing to do.

Chapter 2 | Identity

The ACE model aims to support you in aligning other people through developing clarity in yourself and others. So why speak about identity? Clarity is a subjective experience. What is clear to you might be unclear to others. Thus, your capacity to clarify anything with others lies in your ability to understand them. To understand how you differ from others, you have to get into the habit of understanding yourself. Hence, the ACE model starts with you and your identity.

Identity in the ACE model pertains to the concept of self-discovery. When it comes to yourself, do you know you? What are the experiences that shaped you to be the unique person you are? What motivations do you have? What are your values? What are your needs? What are your beliefs? What triggers your negative feelings? What are the default lenses you wake up with every day that inform the way you perceive this world?

It is easy to give offhand answers to the questions asked above. After all, you have been living in your body since you were born, and you hope that you know yourself. But this is the truth: only a few people know who they are and take responsibility for who they are. These are hard questions. For most, getting to know themselves is a life-long journey.

Leaders who invest in self-discovery often find that they can lead better. They are able to lead authentically, manage their emotional state, exhibit resilience, inspire trust, and create a feeling of safety in others. As flight attendants say, "put your own mask on first." You can only care for other's needs after you have cared for your own needs. How can you care for your needs if you are unaware of what they are?

Every organization witnesses a period of drastic change that cuts across products, services, or initiatives. This is especially true if you are leading a transformation or you are in the middle of a crisis of some sort. During these uncertain times, leaders have two key roles: support their people and translate the organizational perspective. Hence, during uncertain times, it is even more vital to translate well the change vision into a viable approach that can be used to get the best out of the situation. However, it is impossible to translate something you do not understand yourself.

To understand the concept of identity in the ACE model, let's use Ellen as an example.

Case Study: Ellen's Story

Ellen is a young, driven product manager at a small, fast-growing startup. She started as an engineer when the startup was in its early days and grew up with the company. Now she leads the core product development team.

This is her first leadership role. After Ellen continuously demonstrated her expertise in her domain, the startup founders had no doubt that she deserved this job. It's the natural next step. She has great problem-solving skills and the ability to deliver high-quality results. Her dedication and extremely hard-working attitude make her work shine, and she is recognized by many people at the company and beyond. She has earned the trust of her managers, stakeholders, and team. After all, she was the best on her team, and she wanted to lead. Thus, she was ecstatic to accept the promotion and her new leadership role.

Ellen's Context

Ellen recognized the fact that as the company grew, it was more important to bring in more diverse people and delegate more work to others. Her close relationship with the founders enabled her to be involved in strategic discussions that were usually kept behind closed doors. Therefore, she knew that their investors were encouraging the founders to grow faster than the company was prepared for. Also, she knew that the founders were naturally ambitious and optimistic. Hence, they committed to their investors more than they could realistically deliver with the existing team they had. The company needed to take a leap of faith that they would find the right experts to hire moving forward. Ellen anticipated a significant shake-up as they struggled to scale up their engineering team. From her perspective, it was going to be mission impossible.

Simply put, Ellen was tasked with leading a scale-up of a product development team in a startup that was moving to the next stage in the company lifecycle. The founders made a commitment to their investors without fully grasping the current state of the team. The team lacked the expertise in-house. Hence, Ellen struggled to imagine both what success could look like and a path to get there. She understood roughly that there was a need to hire new people and the need to potentially change the team structure and work process as the team grew. Also, she suddenly had more work than she could handle herself.

What Happened?

To support managing a larger number of projects and engineers, Ellen's manager, one of the founders, hired a junior product manager to support Ellen. From his perspective, his job was simple. Ellen would "just" delegate.

As the leader and more senior product manager, Ellen was the primary lead. The junior product manager would support her by working on topics Ellen had no time for. The founders' thinking was that not everything Ellen worked on required a high skill. Hence, it should be a simple task to delegate the simpler work to someone junior and oversee their work. Later, as the team grew enough to split into two subteams, she could split the work so each one would focus on one team. Therefore, the junior product manager was hired not too long after Ellen was officially the new team lead. As such, she had barely adapted to her new responsibilities as a team lead when she had to lead one more person, someone junior to her, to share those new responsibilities.

Here Ellen was, new in her role, struggling to understand the next steps and tasked to lead a new junior person.

Meet Sarah

Filled with enthusiasm and confidence, the junior product manager, Sarah, started her work on Ellen's team. Sarah is a quick learner and very ambitious. She loves learning, exploring, and brainstorming new ideas to do the same thing differently.

Within a month, she asked to take over more responsibilities while not yet mastering the current responsibilities on her plate. In less than six months, she suggested changing the whole process and tools the engineers were using to support the growth of the team while not fully grasping the existing processes and tools they had in place. From Ellen's perspective, Sarah seemed to be much more interested in starting new things than following them through. Her lack of consistency made Ellen uncomfortable. So, when Sarah offered to change the process again, Ellen was puzzled and immediately refused. "You are not thinking things through," she said. Every change Sarah introduced seemed simple but required changing many different things Sarah was unaware of. "This is how we do things here," Ellen said. "It works."

Meet Ellen's Team and Their Work Process

Ellen's team was a typical growing team. When Sarah joined, they were at 15 team members. Their work, which was once interconnected, was now separated. However, in Ellen's opinion, it was not separate enough to justify two teams.

Ellen's team followed one of the known agile frameworks called Scrum. Every two weeks, the team got together to plan what the team was going to work on. Ellen, as the product manager, was prepared with a list of items ordered by priority called a backlog.

Each item on the backlog was selected with a key person in mind. Ellen justified this due to the team being highly specialized. The consequence was that collaboration between team members was rare.

Leadership in a Time of Continuous Technological Change

In the meeting, Ellen explained each item to each individual. Then each member of the team estimated how many days or hours they thought their items would take. The meeting could take two to four hours. Everyone remained in the room, checking their emails and phone until their turn arrived.

By the end of the two weeks, Ellen would check the status of each item. If the item was not completed, they would usually continue to work on it for the next two weeks. The done items would get approved by Ellen and released to the customers. Sometimes, the team demoed new functionality to the founders.

The backlog was never empty because new items were always more important than older items. Frequently, important issues that could not be resolved immediately disappeared from sight and became hidden at the bottom of the list. They became problems at the most uncomfortable times, when it was too late to fix them. As a result, the team wasted a lot of time planning the work for items that were never prioritized, not to mention the long planning meeting where they reviewed all of those items every two weeks. The team was frustrated.

Let's stop right here. What is so important about how Ellen's team works? The way the team works is often a reflection of how the leader thinks. I can talk about agility all day long. I can teach you all the frameworks I know. Moreover, I can customize them to fit exactly what you want or think you want. However, agility, this speed and the flexibility to respond to change smarter and better, starts with you. What in Ellen's identity as a leader contributed to the team working this way? Take a moment and reflect on this.

Contrast: Ellen vs. Sarah

Ellen saw no problem with the current process. It worked. It was hard and it required her to work long hours. However, the team was frustrated. They rarely finished everything they committed to. Nonetheless, it was good enough for Ellen. It was not good enough for Sarah.

To change the process, Sarah suggested splitting the team into two multi-functional subteams so they could work more collaboratively and reduce the time of the planning. She discussed it with the whole team. Considering the further growth and anticipated workload, they were going to waste days of valuable engineering capacity if they continued to work in the current process where people were so specialized.

See the difference?

- Ellen is extremely detailed-oriented, so she struggles to see two separate workstreams. Hence, she prefers the team to remain as one. Sarah is an intuitive, abstract thinker so she is able to see two workstreams.

- Ellen is not concerned about collaboration. She values people working as individuals in their expertise. Sarah wants people to work together.

- Ellen focuses on being thorough. She prefers to take her time to go through each item with the whole team to ensure that everyone understands everything. She focuses on each person and their full understanding of what they should work on and how they should do the work. Sarah focuses on efficiency. She wants to reduce the planning time. Also, she thinks people can figure out the details by themselves later.

- Ellen is present in the now. She prefers to keep things the way they are because "it works" rather than challenging that. Sarah is thinking about the future. She wants to prepare for what is next rather than be satisfied with the now.

Each of these differences relates to the identity of Ellen and Sarah in the context of their work. They were both product managers. It was the same organization. They both focused on leading the same product development team. The results they strived for were the same results. Scrum, as a framework, would remain the same in either Ellen's or Sarah's implementation. The difference here is the mindset. Mindset is highly influenced by identity. Changing it often requires changing ourselves.

Whom do you relate to more? Hard-working, consistent, and result-oriented Ellen? Exploratory, innovative, and learning-oriented Sarah? Maybe somewhere in the middle? Wait, who is right here? Are you leaning towards Ellen? Are you leaning towards Sarah? Did you make your choice based on the person you relate to or the state of the team I described? Maybe you judged it based on what you would have done in that situation. Maybe you struggle to understand Ellen. How blind is she to the lack of efficiency in her team? Maybe she reminds you of this horrible boss you had once who wanted to control everything. Maybe you are Ellen?

When Different Identities Clash

Back to Ellen and Sarah. There was a decision to be made. Without discussing it with anyone, Ellen decided not to change anything. She decided that they should keep the process as it was and not split the team. "They are not ready," she thought. "Sarah is not ready to lead them either."

A few months passed quickly. As a growing startup, the team grew faster than one person could handle. Suddenly, Ellen had over 30 direct reports. The process became more complex. The planning meetings got even longer, sometimes exceeding the day.

With every new member, Sarah became more impatient. She continuously expected the team to break into two subteams. Moreover, at some point, they had enough people to break the team into four subteams. However, that never happened.

Since it was each person to their own, individuals started more topics than they could finish as a team. Therefore, before every delivery date, the team would work extremely long hours to help each other out as they realized the collaborative work was never prioritized.

While Ellen felt grateful to have such a dedicated team, Sarah saw it as mismanagement, carelessness, and unfairness. Ellen truly believed everyone knew what they should be working on and what the priorities were. She was unaware that collaborative tasks were pushed to the last minute because of her individualistic planning approach, assigning individual tasks to individual people without fully considering the outcome. Those long hours were unavoidable.

Less than a year in, Sarah left the company. She justified it by saying that she had no room to do the work she was hired to do and that she was promised her own team. Ellen was perplexed because she dealt with the biggest challenge of her career and her only help quit on her.

Deep Dive
What Was the Challenge?

The perceived organizational challenge was scaling up the engineering team and accelerating the delivery for 30-plus engineers working on the same product. The growth stage of the company required scaling the number of people as well as the number of topics the team was handling. Hence, more work, new types of work, and many new people in a structure that was not designed for any of that.

Was that Ellen's challenge? In many ways, yes. If I focus on the story, sure. Nevertheless, that was not her challenge as a person. It was the first obvious problem everyone in the company faced, not only Ellen.

Ellen struggled to adapt her working style to accommodate such a large team. It meant shifting from the day-to-day tactical work to more strategic and operational work. It meant shifting from focusing on each individual to focusing on coordination and prioritization. It meant taking a step back, looking at things from a higher-level perspective, and providing people with context rather than instructions.

Ellen also struggled to motivate Sarah to grow in her role and enable her engineers to become more self-driven. She was new in her leadership role. It was the first time she was responsible for people instead of a deliverable. No, this transition is not straightforward.

What Was the Root Cause?

Ideally, a great leader should be able to build a strong team of great thinkers who can solve their problems, take initiative, and challenge unrealistic expectations. Most people are not born that way. Most people start their leadership role without understanding what capabilities they need to do it, without knowing what they are naturally great at and what they need to adapt to perform better in such a role.

It's too easy to jump to the conclusion that Ellen is a bad leader. You are likely to think this way if you relate to Sarah or if you had managers that remind you of Ellen. Maybe you think that Ellen should have listened to Sarah and implemented some of the changes together. There is no guarantee that the situation would have evolved more positively if they changed the process.

There are more perspectives you could take. You might support Sarah for running away or support Ellen for doing the best she could. Ultimately, you can say that Ellen might not have the right people on her team or maybe the founders were just too ambitious. Then it is neither about Ellen nor Sarah.

I can go on and on. I love speculating, thinking about why it happened, holding some people accountable, rationalizing, and problem-solving the situation. However, this chapter is about identity. So, how would it help Ellen?

To understand the root cause of this, you first need to understand Ellen. What stops Ellen from becoming the great leader she has the capacity to become? What is the problem Ellen is facing here that only Ellen can solve? She is unaware of her identity. Thus, she is unaware of how her behavior impacted her team and their capacity to deliver results.

People struggle to see how they impact others when they are blind to who they are. Self-awareness is like any other skill. The less self-aware you are, the more you can fall to the Dunning–Kruger effect where you overestimate your ability to reflect on yourself and others. Nonetheless, this is not a skill one learns at school, at least not where I studied.

Leadership requires a level of self-awareness that is hard to quantify. Foundational understanding of who you are is a great start. Again, it's hard to quantify. Thus, let me revise what I mean in saying that Ellen is unaware of her identity.

Who Is Ellen?

As committed, ambitious, and results-oriented as she is, she demanded everyone to align with her needs and beliefs. If you ask Ellen, she never demanded that. She never intended that. Nevertheless, the results were that she became a change blocker rather than a change enabler. Why?

Who is Ellen in this situation? Ellen loves certainty, knowing what is going to happen, when it's going to happen, and by whom. During her career, she was acknowledged for her ability to plan ahead, set up realistic trackable goals, and follow through on those goals, meeting commitments of scope and time, even if it meant working extra hours before major deadlines. To her, it only meant she was hard working and disciplined. Remember, that behavior gained her the trust and respect of her managers, stakeholders, and team.

As a team lead, Ellen was the one planning for her team to ensure that everyone was working on the right things. This is a common behavior of new managers who share Ellen's traits: they ensure the work is done right and by their standards. It is a way to control when one is used to work independently or when one does not trust their team yet.

Should Ellen Be a Leader?

Unknowingly, managers, like Ellen, take full ownership and responsibility for the work of their team. Not the people, but the work they deliver. There is a difference. When the manager controls for everything their team works on, including how they do it, individuals in the team who are not comfortable with this approach may quit. The ones who remain will follow orders.

You cannot fix this with a process or a framework. Ellen's team implemented Scrum, a framework where the whole team should own the delivery. However, the only person in control was Ellen.

It is too simplistic to say that Ellen should have just let go of the responsibility. This is not who Ellen is. She is ambitious and seeks certainty, control, and power. Thus, she gravitates towards leadership roles, but she needs to either find the right environment for her authoritarian leadership style or learn to adapt her leadership style as required in her organization.

Was It All About Ellen's Identity and the Role Fit?

It is not only a matter of a fit to a role. Some people fall into leadership roles and swim; some do not. While Ellen's lack of awareness of her identity contributed to this situation, there are three important factors to keep in mind.

Context matters. Let's not forget that the same behaviors that got Ellen promoted are not the same behaviors she needed to lead. It is not a common practice in many organizations to support individuals interested in leadership roles to understand what behaviors they will have to change to become a successful leader. It is not uncommon for people to switch from leadership roles back to expert roles. It is also not uncommon for leaders to be successful in one environment and fail in others.

No matter what your identity is, it needs to be a conscious choice to learn the capabilities that you are missing. Sure, I can blame Ellen for not being able to step into her leadership role by adapting to the new style and delegating. I can say that she is not a fit for such a role in her organization. However, I feel that this is unfair. The expectations were unclear. She could learn this. Training is the first step. Then, depending on the person's starting point, increasing self-awareness and the ability to accommodate diverse people can take from days to years. We all have the capacity to do these things. Nevertheless, it is important to acknowledge that not everyone is willing to do so. Hence, it needs to be a conscious choice.

Lastly, understanding things rationally does not mean you can or would do something. Rationally, Ellen understood that the company was growing and it would be impossible to deliver the work they committed to without scaling the engineering department. Unconsciously, she lacked the trust that her current team could deliver good enough results without her supervision. It was not even about the team. Some people are naturally more suspicious of others. Thus, Ellen understood rationally that she would struggle to scale the team on her own. She accepted hiring Sarah. Moreover, she agreed to the plan to split the team once the junior product manager, Sarah, was onboard. What Ellen was not aware of was how the feeling of losing control would paralyze her. Delegating work meant not having the certainty that things would happen exactly the way she wanted. It meant Ellen was no longer the one to prioritize and explain every item to the people on the team. It required trusting others to be responsible while she was accountable. As an individual team contributor, your level of trust might not necessarily influence your results. However, how can Ellen enable Sarah to take over control without delegating to her or trusting her? It is not enough to know what you need to do.

Why Can't It Be a Process Problem?

Let's take a hypothetical scenario where Ellen trusted Sarah. She would have enabled her to implement the desired changes in the process. They could have done it together.

Would Ellen step back and allow the team to plan their work collaboratively with Sarah? Merely the idea of not overseeing half of the engineers was so stressful for Ellen to an extent that she blocked it. If Sarah implemented an idea and it failed for any reason, Ellen would have seen that as validation that she was right. It would have been even harder to convince her to let go of control in the future. Hence, before you run to problem-solve the process, start with an inner solve.

In the beginning of this chapter, I explained that when a person's identity is in alignment with their environment or is perceived positively by their environment, they are in harmony. People partially create their environment at work by selecting frameworks, processes, best practices, and standards to operate by. When they are not aligned with the leader, it is common to observe dysfunctional implementations, such as how Ellen's team implemented Scrum.

Case Study Conclusion

Ultimately, the root cause of this situation was Ellen's lack of identity awareness. She would benefit from a better understanding of who she is, what her needs are, what she values, what triggers her, and what motivates her. If she does not know, she can't make the conscious choice to improve. She will continue to struggle with supporting her company growth. Also, she will struggle to work with different types of people such as Sarah.

As a leader, knowing yourself is crucial. It does not mean you need to have a solution right away for everything you struggle to do, but you do need to know how you perceive the world and make decisions.

When you understand yourself better, you can decide whether you want to change yourself and your situation or accept the current state of things so you can focus on challenges that matter to you more. When the problem is clear, people can support you in finding a solution that works.

Kick-Start Your Self-Discovery

The ACE model is all about clarity and alignment. Hence, I want to ensure that I am clear first. Self-discovery is a process of discovering who you are so that you can get to know yourself better. Hence, your identity. It is a life-long journey that is easier for some and extremely hard for others. It requires a lot of inner work. Do not shy away from asking for support with this. I acknowledge you for even considering this. Do not force it; see what works for you.

Remember that as a leader, you are also going through the same change as the organization. Have some empathy towards yourself first. You are not going to be everything to everyone, nor you should be. Also, remember that not every environment fits every person, and nor should it. Hence, if this self-discovery process leads you to consider a role or organization transition, you might want to consider speaking to a career coach. This helped me changing to something that fits me. I had my fair share of career transitions as a result of this. While I learned a lot from each role, I would lie if I said they were all successful. Alternatively, ask yourself how much you are willing to become the person that fits a change, a role, or an environment. Some people are highly adaptable. Lastly, if you are supporting a change that was decided from above you, take the time to explore what it means for you, not just to the organization.

Since this book focuses on uncertain times, I do recognize that change during such times might lead to a lot of unknowns. There will be times when you will struggle to recognize what you do not know. There will be times when you face a challenge, like Ellen did, and you will not be able to identify if you are blocking it. Self-awareness is the foundation to start exploring this. Understand who you are, how you impact others, how others impact you. By asking yourself questions, by including others in your problem-solving process, you are increasing the likelihood of being an enabler rather than a blocker. Where to start? Self-discovery is the first step.

Personality Assessments

The simplest way to gain initial data and insight into your personality is to take a personality assessment. This step helps to get to know your preferences through answering specific type of questions honestly. The key here is *honestly*.

There are many types of personality assessments depending on the information you are seeking about yourself. Assessments such as the Big Five, DISC, or Myers-Briggs Type Indicator (MBTI) focus more on the natural traits and preferences of the individual, interaction with others, and gathering or organizing information. Alternatively, assessments such as StrengthsFinder focus on identifying your natural talents, which can then be developed to strengths. These are only a few options. There are many more personality assessments. Thus, it is highly recommended to seek a professional such as a career, leadership, or life coach before going through these assessments. Make sure that they are certified on one or more of those type of assessments.

Recommended or not, do not let finding a coach to be your reason for not taking an assessment. Again, technology comes to the rescue here as you can utilize a personality test online to discover a bit about who you are. With thousands of test platforms and apps, you can gain an initial awareness of how you see yourself, which becomes helpful to you both on the job and in your personal life. Most of these personality tests are free.

Be mindful that taking a personality assessment is one step in getting to know yourself. It is a mirror to how you see yourself right now. It can change when you change. It can also change when you are in a bad mood. Why? Most personality assessments are based on self-evaluation. Thus, they are highly subjective. To gain a complete picture, it is highly recommended to ask people who know you and work with you to provide feedback on the results or even fill in the assessments for you.

Some of the assessments suggest that your personality is fixed and you should stick to who you are. Others suggest that you can change and improve. Ultimately, the purpose of assessments is to give you an insight so you can have a language to describe your preferences. You are in control of your life and it is your choice whether you want to follow the results or change them. Hence, my personal recommendation is to not follow any framework religiously. Yes, some traits are fixed. Nevertheless, we can often change a lot more than we think when we believe that we can change.

What worked for me? When I started my self-discovery process, I wanted to religiously believe in one or two frameworks. However, I must admit that I knew myself so little that I could not relate to any of the profiles I received. What helped me the most was gathering anonymous feedback from people who know me. Yes, anonymous. I tried to have honest conversations with people face to face but I was too sensitive and judgmental to listen. Hence, people walked on eggshells around me, telling me what I wanted to hear rather than the truth. The first time I accepted who I am was when I saw my written anonymous feedback. It was the first time I realized many people saw me in a very similar way, regardless of who they were. It was hard for me to argue with that.

Visualization

Identity can also be revealed via exploration and visualization of an ideal day, future, or an activity. Become more mindful and curious about your own experiences. What are the things you love so much that when you do them you lose track of time? What does an ideal day look like? If you could be anything you want to be, what would it look like? What are the ideas you promote in your mind? What are the things you can't stop thinking about? The answers you provide to these questions will give you insights to who you are and what you value and like.

Visualization has always been a powerful tool for self-identification because a person is the combination of their feelings, thoughts, behavior, and experiences. When we think about an ideal day, we go through multiple situations and decide how we would like them to be. When those ideal or important situations happen differently in reality, we may identify a potential trigger. If reality is better than our imagination, we are happy. If reality is worse than our imagination, we are not so happy. What we consider as better or worse is also subjective.

We respond to situations based on our dominant thought pattern. For example, Ellen's ideal day is well planned so she knows what to do, when to do it, and how to do it. When she needs to live in the moment and make decisions based on unknown information, she is triggered. Sarah, on the other hand, loves to explore new things every single day. The worst thing for her would be every day looking the same.

Be mindful that some people struggle with visualization. Hence, if you can't visualize in your mind, try writing things down, drawing a picture, or sharing your thoughts with a coach or a friend who can support you in exploring further. As I struggle to visualize myself, I often try the fast-forward method where I ask myself: what if I woke up tomorrow and the decisions for the day were already made, and everything turned out exactly the way I wanted, what would that look like? I write it all down in past tense as if I am summarizing what happened that day. A visual friend of mine does the same thing with pictures. She collects or draws the symbols in her mind. Sometimes, we do it together, so she finds the pictures to visualize my story. If you can see it, you can imagine it. If you can imagine it, you can do it.

Mindfulness

Mindfulness refers to awareness; it is a state in which a person is aware of their present state while acknowledging their feelings, thoughts, and bodily sensations.

Being mindful makes it possible for you to identify what makes you tick, the things you don't like, and the things you are passionate about. You can practice mindfulness by taking time each day to analyze your feelings and thoughts about the issues you dealt with during the day. Every emotion or thought is an invite for an exploration.

Ask yourself during the day "how do I feel?" or "how do I think?" When feelings are hard to grasp, even the answer "good" or "bad," is a great start. With a thought, I try to write down the story I tell myself. For example, a manager once told me that he expected more from me. I felt bad and thought "I am not good enough to do this job." Don't judge those feelings. Feelings are just feelings. A manifestation of the story you tell yourself. Instead, focus on building the discipline to write them down as frequently as possible and review them later in the day. If possible, get the assistance of a friend, partner, or a coach when you identify certain situations where you feel or think in a certain way that is misaligned with your expectations or wishes. In my example, I definitely did not want to go into victim mode of not being good enough. I wanted to respond with "tell me more" so I could do better next time.

When you get to know who you are, it becomes easier to understand how certain changes, concepts, and solutions will impact you. Thus, it will become easier to recognize which ones you will naturally support and which ones you will struggle with yourself. Knowing this will enable you to better communicate expectations with your leadership and team. It will also ensure you have the support system you need.

Case Study Epilogue

Ellen took some time off to explore what was standing between her and supporting the company growth. She discovered that she naturally seeks certainty and not having control over the whole team was extremely stressful for her. At the same time, it was unrealistic to expect that. Her need of control took away the opportunity for others to think and own their work themselves. As the team grew, it would be impossible for her to be the sole owner because it's already diminishing the performance of her team. Therefore, she walked back to her team and expressed that. The team was surprised and acknowledged her for her vulnerability. Together, they found a middle ground where Ellen remains involved in all planning conversations until she is ready to let the team take over. Then, they proactively involve her when they need her input and inform her of the decisions they make. She is still able to intervene early if needed. Furthermore, this willingness of the team to help, support, and include Ellen to accommodate her journey to become less controlling enabled Ellen to gain trust in her team. When she needed them, they were there for her.

Ellen wants to be an inclusive leader and support her team to grow and deliver the best quality outcome they can. Most controlling leaders out there do. Also, in some situations, being more controlling is needed, such as when the team is very junior or at the time of crisis where decisions cannot wait.

Realizing this is important. Ellen's tendency is neither good nor bad. It was just not the right tool for the type of team she had, the environment she operated in, and the change her organization wanted to drive. Now that she is more aware of her own fears and needs, she is able to express them to her team. This is an important step. No one can know what is in your head and heart if you don't share it.

Conclusion

In this chapter, you got to know Ellen. I shared her story of supporting her company from a small startup with one product development team to a larger company with multiple product development teams.

You also got to know Sarah. They are very different people with very different identities. As a result, they also had a different approach to their role. However, Ellen's seniority influenced the decision-making. Thus, Sarah ideas were never implemented, and she left the company.

Ellen's team suffered from the situation. Ellen's attitude towards the work was not aligned with what the team needed to be successful during those uncertain scaling times.

We dove deeper into what happened there. We analyzed the root cause, asked whether it was a role fit, and discussed the alternative that it was a process problem. The conclusion was that Ellen's lack of identity awareness played an important role in the struggles Sarah and the team faced. It hindered the change the company was undertaking.

If you take one thing from this chapter, I hope that it is that you must start with yourself. As a leader, you can take the same steps Ellen took. Ask yourself "Who am I?" Then, with the aid of the self-discovery approaches, you can communicate better what you need to operate as the best version of yourself. Help other people to align with you or find a middle ground. Own it. Realize how you can grow, where you are movable or fixed, so that if you want, you can become an even better version of yourself.

Questions to get you started:

1. Who are you?
2. What motivates you?
3. What makes you unique?
4. What do you value?
5. What are your needs?
6. What do you believe to be true above all?
7. How do you see yourself in comparison to others?
8. What type of a person are you at work?
9. What type of a person are you at home?
10. What experiences shaped you to be the unique person you are?

11. What situations trigger your negative feelings?
12. How do you feel about your current leadership position?
13. What does leadership mean to you?
14. What kind of a leader would you like to be?
15. What strengths do you bring to the table?

Ready for the next phase of the ACE model? We are going to dive into the concept of emancipation: flip to the next chapter and discover more.

CHAPTER 3

Emancipation

Create Space for Your People to Contribute

Emancipation refers to the aspect of being set free from legal, social, or political restrictions. It is about gaining freedom in a context where one is not in control. For example, children are emancipated from their parents when they come of age. However, depending on the relationship, emancipation can be a natural process or a painful one.

Concerning organizations, control is an ongoing discussion. Every company operates with an organizational structure that defines who is in control. Someone must be in charge and make the final decision. Moreover, flat hierarchies are rarely flat. Some people raise themselves to positions of power while others prefer not to. Like it or not, discussions about structure, control, and responsibilities are critical for organizations to operate effectively. We will dive deeper into this topic in Chapter 4.

In this context, freedom, the antonym of control, stands for *what I can do without asking for permission*. Employees in most organizations rely on a boss to decide what they can do. The boss, the person in charge, approves vacation days, provides feedback on the work you did, and can terminate your employment.

© Bar Schwartz 2020
B. Schwartz, *Leadership in a Time of Continuous Technological Change*,
https://doi.org/10.1007/978-1-4842-6300-6_3

Chapter 3 | Emancipation

When your boss is completely hands-off, it feels like no one cares what you do. Also, if there are no consequences for bad behavior, it is equal to giving implicit permission to act that way. On the other hand, if your boss is too hands-on, it feels like micromanagement. That you should never decide anything on your own. At some point, people either leave or stop thinking, waiting for their manager to tell them what to do.

Emancipating your people means that you intentionally give them freedom. However, that freedom has boundaries around it so it does not mean they can do whatever they want to do. It means that within certain boundaries, they have the freedom to make decisions.

Leaders typically have all the control. Inclusive leaders delegate decision-making power to the people they lead to seek solutions that fit their unique context. However, in traditional hierarchies, where leaders are expected to have all the answers, leaders feel pressured to come up with all the solutions. This is stressful and centralizes the risk on one person's judgment. Often, it pushes leaders to seek solutions externally rather than utilizing the people they have, so it also outsources the judgement.

This is not a knock against consultants and external support. The job of a consultant is to offer an advice when you lack the capabilities in-house to solve the problem. My opinion is that many leaders rush to assume their people lack capabilities, while in practice, they never gave them the chance. Therefore, in this chapter, I invite you to imagine a reality where your people can make decisions that are aligned with the needs. Where you are not the only decision-maker and not everything is on your shoulders. Where you are not obligated to know all the answers. You emancipate your people from having to rely on you and only you. Then they can support you in a way that fits them.

This is a scary concept for many leaders, people like myself, who like being in control and are afraid of others not meeting quality standards if they are not there to hold their hands. In many ways, how is this different from over-parenting? We work with adults so there is no real need to hold their hands unless they say they want us to.

Being a control freak, my belief behind it is that I need to be the hero who figures it out for them. That they will not be able to figure it out by themselves. Thus, every time they are unable to figure it out by themselves, I confirm this belief and jump right in. I convince myself further that they will fail without me. While there are many ways to build your self-esteem, mine was by assuming that everyone needs my rescue. I need to save the day. However, it was just my perception that people need saving.

The problem with this way of thinking is that it is not scalable and, ultimately, it is not true. The argument here is that leaders are so much better than everyone else in their job so only they can decide what is good for everyone.

If that's true, why do so many leaders say that they want to hire people who are better than them, people who bring new capabilities to the table, someone to learn from? Why spend millions in attracting, hiring, and retaining top talents with experience just to tell them what to do? Based on this logic, only the CEO should make all the decisions. If you are not the CEO, your boss knows better than you. How do you feel about that?

I get it. It is easier said than done. Your people will fail without you. Some of them are weak, incapable, junior, not ready. I've heard it all, from my own head and others' mouths. A lot of good intentions have driven great people out of many organizations. There are many excuses that controlling leaders tell themselves in order to keep thinking that they know best. If you do this, whatever the excuse you have, the underlying root of that is that you lack trust in your people.

Take a deep breath. Trust is hard. Building it is harder. When trust is gone, relationships of any type fall apart. It is also hard to recognize when trust is not there. I thought I was a very trusting person until I realized that I trusted no one.

Imagine a stranger coming to you on the street and asking for your bank account details. How likely are you to give them to this stranger? Very few people would say yes to this. Now, imagine that a friend of yours asks you for your bank account details. You know this person. How likely are you to do it now? Of course, it depends on the friend and the situation. However, I think many more people would say yes to that. It is the same for people you work with. If your team is a group of strangers, it is very unlikely that you will trust them. However, if you know each other for months or years and you still trust them as much as you would a stranger, something is wrong. Be honest with yourself: how much have you invested in getting to know them?

To summarize, emancipation is the idea that you can trust the people you lead to contribute to decision-making by being enabled to make decisions within their boundaries. In other words, when people are given the freedom to decide, they are capable of making the right decision.

Emancipation, however, is not a given. I wish it was as simple as telling people that they are responsible for something and fade away. The best expert is unable to make the right decision without understanding the context and boundaries of their decision space. Too many leaders struggle to explain this. Are you able to make decisions with no information? George Harrison sang it best in his song *Any Road*, "If you don't know where you're going, any road will take you there."

Emancipation is not just trusting the decisions of others; it is creating the space for them to decide correctly because they understand the context they operate in and they know their job better than you, so they can make their own decisions, which will be better than yours. Let them do what they know, their way. This is important for their inner alignment with their work and the organization.

Remember that you are not alone in this. You have a team. You lead people. When faced with a problem, many leaders feel that they are the sole person responsible to solve the problem. They misunderstand the whole accountability vs. responsibility game. I understand that traditionally leaders were responsible for results. However, you are not in control of the results because you are not doing all the work yourself. In the work context, you are in control of people. They are in control of the results. Hence, you are responsible for the people, and they are responsible for their work. Together, you are accountable for the organizational results, the outcomes. Your job is to emancipate them by clarifying what those outcomes are.

To understand the concept of emancipation in the ACE model, let's use Nick as an example.

Case Study: Nick's Story

As Nick stepped into the company's headquarters for the quarterly C-level meeting, he sensed the tensed atmosphere and immediately knew something was off. He hurried quickly to the meeting room, and the faces of others present at the meeting confirmed his suspicion.

The CEO started the meeting by stating his displeasure at the reduction in sales. As a product manager, Nick recently realized that he was as accountable for sales as the sales team. However, at that moment, he felt completely helpless when he was asked to explain the reasons for the bad sales. He was not a salesperson and he was not part of the sales team.

Nick recently stepped into the product manager role. With an engineering background, his knowledge in coming up with a good sales strategy or analyzing how a conversion funnel performs was close to zero. From his perspective, his role was to build the best product. It never occurred to him that every product must have a target audience who is willing to pay money for it. It must be commercially viable. That it was his responsibility if the product makes no money. Hence, it was not obvious to him that he needed to step up his skill in this domain.

The company was still in its early startup days. Most people Nick worked with, including his team, were junior. It was common for the company to hire people straight out of college. Also, the founders met in college. It was something to be proud of. Hence, Nick felt alone against this new problem.

At the end of the meeting, Nick suggested that the company hire a sales consultant to proffer solutions. He thought it would be faster to bring in an expert to solve the problem since clearly they lacked the capability in-house.

This opinion was not shared by all. However, after long discussions, the CEO agreed, and the consultant was hired. Ultimately, it was Nick's decision and his budget. If he believed that no one on his team could support him, he must know best.

The consultant started a couple of weeks after the meeting. Shortly after, a new strategy was implemented, and everyone hoped that things would get better. However, when sales were even worse a few months later, Nick was puzzled. It should have worked. They hired an expert.

The consultant was truly best in class with a lot of experience from different companies. They suggested many great organic methods that should have set the products rising through the market and dominating the sector. Everything the consultant offered was best practices. Nothing was new. Also, the person was smart, had a great reputation, and seemed to know what they were doing. So, how come nothing worked? Why did the numbers go down?

Because Nick did not emancipate his team.

Deep Dive
What Was the Challenge?

The perceived challenge was the lack of in-house ability to create a sales strategy. Nick, as the leader, lacked this experience. He also assumed that his team lacked this experience as well due to their perceived seniority level. Lastly, he assumed no one in the whole organization could solve that problem. Therefore, he made a decision to not engage his team or anyone else in the organization in solving the sales problem.

Very often, the first problem is not the real problem. Many people have a weird expectation that they can or must figure things out right away. Leaders, due to the expectation of having all of the answers, often think that problems are clear. Hence, they avoid spending time talking about the problem. They focus, instead, on bringing solutions. Thus, Nick did not understand the sales problem. However, he felt pressured to offer a fast solution.

What Was the Root Cause?

Nick is not a salesperson. He is an engineer who took over the role of the product manager. Thus, he felt he should know everything about the product, including how to sell it. However, in this specific case, he had no idea what was causing their sales to drop because he lacked the skills and knowledge.

Ideally, when someone moves into a new role, they also receive the support required to transition to this role. This might require training, coaching, and job shadowing. Nick's manager should have clarified with Nick what was missing and support him in developing a plan on how to compensate for any missing capabilities. We will discuss this further in Chapter 4.

Nick had no trust in his capabilities to solve this sales problem. However, he also lacked the trust in his team to solve it. His reason was their seniority level. Where does that belief come from? Remember identity? Who is Nick as a person? What in people's seniority level impacted Nick's decision to not ask for their contribution?

Building on that belief, for months Nick told his team to sell the product in a very specific way, the consultant's best practice way. The external consultant Nick hired worked separately from the team using Nick's opinions and data they gathered from the company tools. Critically, the consultant never met people in the company and was not familiar with the current capabilities of the team or organization.

When the salespeople received the new strategy, they were displaced. They knew the problems they faced when trying to sell the product. The suggested approach was very different from the way they were working and none of it addressed their challenges.

Nick's product development team had to create many new features to support the salespeople in implementing the strategy. These features were not aligned with the current way the product was built and designed. These features were additional work but nothing was dropped, stopped, or delayed because of these features. Moreover, it was unclear what made these changes necessary. Nonetheless, the team felt that they had no choice but to implement them.

When the sales worsened, no one was surprised. They said, "The new strategy tried to correct things that were already working." In many ways, it falsified Nick's belief that his people could not solve that problem because they were too junior. Of course, we are only able to judge their solutions after those are implemented. The point here is that they were never asked.

Why Couldn't They Hire an Expert?

Consultants are great support when you are sure that you lack the capability in your organization or when you seek a second opinion on a solution you are considering. Ultimately, there is nothing wrong with Nick's decision to hire a consultant. It was the decision to keep everyone else out of the process that created the friction.

I could speculate whether a different expert would have been able to engage the team better and not rely on Nick and data from tools to solve this problem. However, I know that Nick felt very uncomfortable involving more people in this. He thought that it would be a waste of their time because they were not experienced enough. Whenever the consultant wanted to talk to someone, Nick blocked it, suggesting that he knew the answer or that he would speak to the team to save time.

Why Didn't the Salespeople Take It Over?

Lack of transparency is a common challenge in many organizations. Information is shared in meetings. Those meetings are often ineffective due to the wrong people being in the room and lack of facilitation. Thus, the salespeople knew something was wrong with the sales, but they were not part of the discussion of how to improve it.

Another reason was that no one felt accountable. Like many sales teams, each salesperson was responsible for their sales. When they met their numbers, they were happy. When they struggled to meet their numbers for a continuous amount of time, they were let go. Thus, the remaining salespeople focused on selling rather than questioning the product. It was hard enough to keep their numbers up.

Why Didn't Nick's Product Development Team Take It Over?

Nick's team was offended that they were not part of the discussion. They saw Nick as "too full of himself" and "corporate." They took it personally. Listening to some conversations, I discovered that they secretly hoped for the project to fail.

You might get alarmed here. Should we even tolerate such behavior? Remember what I said about being hands-off about behavioral standards? If people are not held accountable for bad behavior, it is the same as permitting it. Also, many leaders create an unhealthy culture in their organization, which causes people to resist rather than support. Emancipation is setting people free. If we over control, people resist, rebel, disengage, or leave.

Case Study Conclusion

The concept of emancipation encourages leaders to shift their mindset from "as the leader I know best" to "the people doing the work know best." It is about believing that people are already empowered, resourceful, and capable of solving their own problems rather than requiring rescue. Our job is to enable them to solve their own problems and not to jump in to save the day with our answers.

Emancipation is the foundation for operating inclusively. When people are involved, they contribute to solving the same problems, achieving the same goals, and exploring the same ideas with you. This increases your chances of creating better results merely by involving more brains.

Teams who have experienced emancipation are also far more aligned. They understand what is going on. They feel accountable for solving the challenges and problems they experience in their day-to-day work. They contribute their own preferences, ideas, and unique experience, not just their execution power. Hence, they see themselves responsible for moving the organization forward.

When you let go of controlling the solution space and telling other people how to solve their problems, they sometimes surprise you with a much more creative way to solve their own problems. As they take more responsibility for their work and develop a sense of accountability for the organization as a whole, you can trust that they will work with you, not against you.

Emancipate Your People

Your team needs three things to achieve emancipation: a purpose, a culture, and a mission. In short,

- Why are we here?
- How should we work together?
- What is the most important thing to achieve right now?

Purpose

Purpose refers to clarifying the reasons why something is done or exists. It is about creating a vivid vision that enables people to imagine a desirable future and themselves in it. In many ways, purpose is why we are here. Purpose is why we care.

Imagine you had a time machine to visit a day in the future. On that day you already achieved the things you are currently working to change. What would it look like to be successful? What would change for your company, your team, and yourself when your current challenge is resolved and your vision is a reality?

Vague Statements Are Unclear

Most companies have a vague mission statement. This is not just unclear; it is also dangerous. When the mission statement is vague, people in the organization tend to see it as a marketing message rather than a guiding compass to where the organization is moving. The easiest things to default to is that *we are here to make money for our executive leaders and shareholders.*

Vague mission statements create a different image for each person that hears them. This image could lead to people working toward a different outcome than what the leadership intended. Clarifying your long-term direction is an important step for emancipation.

Working with many organizations throughout my career in the software industry, most of the leaders I have met considered having a vision as a fluff. They think it's a soft skill, it's not really important, and it's a waste of time compared to coming up with a new strategy, financial model, or dashboard to track key performance indicators. We invest so much energy and time aligning on status, numbers, and tactical tasks. Alignment on a shared vision is as important as aligning on your targets. Where are you going otherwise?

What makes so many people think that this type of work is fluff is the huge confusion out there on the difference between purpose, mission, vision, and all those terms. In this book, *purpose* stands for the reason for our existence and the direction it implies that we are moving towards. You can call it *vision*, *mission*, or *strategy*. What matters is that you understand what you want and why you want it. Then you can also measure it and talk about it clearly. Then it will stop feeling like fluff.

Creating Wealth As a Purpose

You are now familiar with Nick's problem. Let's take a step back and talk about Nick's organization. What is the purpose of Nick's organization? At the time, this was only clear to the CEO. The rest thought it was about creating wealth for the shareholders.

Every organization starts from an idea to advance a certain cause and/or provide value in a certain way to a group of people. Whether the founder is aware of this or steps into it by mistake is a topic for a different book.

As the company grows, the sense of purpose it had in the early stages starts to fade away. New people are hired. They have no shared experiences with the people who have been there since the beginning. For them, the current state of the organization is what is. Nick's organization was no different.

As mentioned, due to the lack of clarity, the default assumption was that they were there to create wealth, to generate money. People at Nick's organization were focused on execution towards revenue and on key performance indicators such as sales volume and conversion rates. They saw their job as moneymakers. Since they lacked the context of what the company was about, they assumed it was nonexistent.

Creating wealth is a valid purpose. If that is your true purpose for your business, make it very clear to your people. Attract people who are motivated by money. Stick to revenue streams that have the highest likelihood to be profitable and cut operational cost whenever possible. Also, reward people with money for success. There's no need to hide behind a vague statement. There are people in the world who are motivated by wealth.

There are also many people who are not motivated by wealth. For them, earning money is a means to maintain a certain lifestyle rather than their purpose. These people will naturally commit less to your organization. They do not care about what you care about.

It is not uncommon and it is very possible that the purpose of your organization is generating wealth. However, I invite you to challenge that. Think about it as the easy answer for those who want to avoid wasting time thinking about it. After all, every organization wants to grow and generate money sustainably. If you are not profitable, how long can you keep paying salaries to all of your great people? However, would you say that generating wealth and earning money gets you out of bed in the morning? Does it inspire you to bring your best self to work?

Creating a Cause

Purpose creates a cause to serve that purpose. It is your unique contribution to the world. It is the change you want to create or the problem you want to solve. It inspires a sense of determination in you as you translate it from a vision into a reality. The people who share the vision will want to work with you because they want to see the same future. The great news is that we all have that purpose! Simon Sinek called it a "why" and suggested that you should start with it.

How will people know what you stand for or what your organization stands for if you never share it with them? Having the language to describe and communicate your purpose is as important as having a purpose. Again, vague statements are unclear and dangerous. Finding the purpose of your organization starts with your personal "why."

Finding your "why" is about seeking patterns in the decisions you make every day as they reflect your values and reasons. One technique to do this is to write down stories of events and decisions that you consider as life turning. They can be positive or negative as long as they had an important impact on your life.

Next, share these stories with a partner, a coach, or a friend, allowing them to ask you exploratory questions that support you to dive deeper into your feelings and reasons. What was important about that decision? What alternatives were available to you? What drove you to choose this one alternative and not the other? What did you value more? As your partner writes the answers down, you may discover patterns in your behavior, feelings, and thoughts. This pattern is often your "why." Your work is one of the ways you live your purpose daily.

Now do the same process for your organization. Gather stories from people who have been there since the beginning. What makes this place what it is? What events transformed the company? Gather success and failure stories. Are there any reoccurring patterns?

I recommend the following authors, who inspired me to understand my purpose: Simon Sinek, whom I mention multiple times in this book, and Cameron Herold. If you are unclear about what makes a purpose so important for an organization, read *Start with Why* by Simon Sinek. If you are seeking to discover your own purpose, read *Find Your Why* by Simon Sinek. If the previous is clear and your challenge is in crafting all of that into a vision, read *Vivid Vision* by Cameron Herold.

Clarifying the Purpose

As a leader, you want every person you lead to be clear on what the purpose is so that they can work toward the same future, even when you are not there to remind them. Therefore, your primary responsibilities as a leader are to know what this purpose is, explicitly communicate the purpose to everyone, and then link everything you do daily to that purpose.

This level of clarity is crucial whether you are a founder, the CEO, a department lead, a team lead, or an individual contributor who wants to lead. The catch is that if you are not the owner and you work for an organization that is not aligned with your purpose, you will struggle to do this. If you choose to remain a leader in such an organization, you will have to align yourself before you can align others.

Nick is a leader at an existing startup, so his role is to translate the CEO's vision to the people he leads. This includes supporting the sales team and his product development team in having clarity on the reasons certain decisions were made. Furthermore, it also includes connecting their contribution to the larger organizational vision. It is particularly important when you are the owner. Then, the company purpose is a reflection of your purpose so ensure you know what it means to you so that you can explain it to others.

Nick's salespeople truly thought the purpose of the company was to generate more wealth to their shareholders. Therefore, the team focused on generating money from new customers with no concrete focus on a target niche.

The team was extremely aggressive with their customers as it enabled them to close deals fast and raise their numbers. For them, the goal was to increase sales as soon as possible at all cost. Remember, those who failed were let go.

Considering the product, many customers realized quickly that the product was not a great fit and requested a refund. That led to the sales decrease. However, no one measured it, so salespeople kept being rewarded for closing deals, completely ignoring the quality of those deals.

When the consultant provided a targeted sales strategy to focus the sales team on what mattered, the team was displeased because they were unclear on the reasons for focusing on that specific customer niche. Their aggressive methods were insufficient with that niche. Hence, they were concerned about their ability to meet their numbers and preferred to stick to the type of customers they knew how to sell to.

The product team developed the product in a silo for their specific target niche. The two niches were not the same. Also, they were unaware of the type of customers the sales team was attempting to sell to. The complaints from those customers were never escalated.

Lastly, lacking clarity meant that it was hard to prioritize. Therefore, all the changes that emerged from the consultant's strategy increased the original roadmap scope. The work was added on top. The deadlines were not adjusted.

When leaders clarify the purpose, people understand better what they are trying to achieve. They understand the purpose behind the decisions. They are clear on the desired customer niche and the problem the organization is aiming to solve. They can work towards the same future instead of driving their agenda or making assumptions.

Nick has to repeat and clarify what the organization was about, even if it is money, so that the sales team, the product development team, or both can adapt. If you are clarifying the purpose of your organization, remember that you can never over-communicate this.

Culture

Culture refers to the social behavior and norms found in human societies, and it is peculiar to every individual. Every organization has a culture. It's the unwritten rules of how people behave, communicate, and work to achieve the company objectives, primarily when no one from the management is looking.

Culture is not who we think we are, it is what we do. When everyone associates culture with values and beliefs, none of it is real if what you say is not what you do. Hence, you may have a written and communicated culture or not. You may believe that culture is created by each leader and influences their team only or not. You may think that your culture is fixed or not. You may argue what your culture is. Ultimately, your culture is what you do so your team culture is what they do (also when you are not looking).

Virtues Influence Culture

There are many resources about organizational culture. The ones I found the most useful in my journey were *The Owner's Manuel for Values at Work* by Pierce J. Howard and *What You Do is Who You Are* by Ben Horowitz. The first book clarified to me what values are, how to assess them, and how to utilize them in a work setting. The second challenged my belief that values create culture. They don't, virtues do. Together, however, these books gave me a comprehensive understanding of the gap between what we believe and what we do. Understanding this is vital for driving any type of change in any aspect of your life.

While I could write a whole book about this topic, I do want to simplify it to the context we are at, aligning on change inclusively during uncertain times. When it comes to culture, every organization is balancing an imaginary triangle. At the apex stands the organizational culture, the unique work culture that represents how people work together to achieve their day-to-day goals. At the base vertexes are the leader's culture and the employee's culture. They represent the culture that individuals bring to the organization. With each individual, the organization slightly changes. With each leader, the change in the culture is larger. This is a normal transition as individual contributors tend to seek the approval of their managers, so they are likely to adopt the behaviors the new leader exhibits. Hence, over time, we are more likely to adapt to the dominant organizational culture unless we have enough people at the top who share our culture. No wonder we all must lead by example.

Why does this happen? Your organization has a unique work culture. However, you as a person, also have a culture. Your culture is driven by the country you came from, how you were raised, and the social rules you learned to follow. Therefore, when you join an organization that shares your values, default behavioral tendencies, and world view, you feel at home. However, if you join an organization that follows a significantly different culture than your own, it feels odd. This is the reason why you can do the exact same role at two different organizations and feel completely different doing it.

Unfortunately, it means that if you are a person of habits who dislikes having to adapt yourself and your behavior, it is going to be critical for you to fully assess the gap between the organizational culture and your culture to ensure you avoid taking a job in an environment that doesn't fit you. It also means that if you are the person who is leading the cultural change, you must ensure that you are not doing this alone or against the executive team's wishes.

The good news is that as long as major values of ours are not being violated, we can adjust to a new culture. We are much more adaptable than we think we are. However, we first must recognize what this culture is. What are the virtues in this organization?

Leadership Behavior Sets Examples

Be mindful that unless you are the owner, founder, CEO, or the person in charge of the cultural change, you are expected to follow the culture the organization aspires to have. It might not be the culture they have right now. When you, as a leader, follow a different culture than the organization aspires to, the people you lead will also follow that culture. This often creates problems and friction between individuals across the company. As mentioned, people look at you to judge what they can or should do.

Following a different culture may lead to people under-evaluating your performance. The most common issue with female leaders in male-dominant corporate environments is the expectation to be decisive. As females, scientifically, tend to be more agreeable than their male counterparts, they tend to agree more and seek harmony and common ground before deciding. In some environments, this is considered weak. Hence, people reporting to that leader may think that she is weak. This is not reality; it is purely perception.

Let's discuss this for a moment. This is a cultural bias. First, not all female leaders are agreeable and not all male leaders are disagreeable. Assuming this may create situations where you hire a female assuming that she will be agreeable. In environments that value consensus over speed, you may blindly hire a female and be surprised that she is very competitive. Second, in consensuses-driven environments, agreeable people will succeed better than disagreeable ones, regardless of gender. If they are men, they may be perceived as weak in more competitive environments. Hence, do not assume culture based on biases. What matters is how much you are at odds with the dominant culture and how much you are willing to adapt yourself to the organization's culture.

Every new person who joins your team will look up to you to learn the culture. Your behavior will define their new norms on what is expected of them. As a leader, you are not just responsible for following the desired organization culture yourself; you are also responsible for teaching others how to follow the same culture.

It is true that you may have a few people who will not be as adaptable. However, as mentioned, lack of alignment with the culture is usually perceived as low performance by others. It will be extremely hard for anyone to behave differently than the rest of the organization. Hence, unless it is your company or your job to change the culture, help your people to adapt or find a different place where they belong.

Clarifying the Culture

Nick worked in corporations for most of his career until he joined the startup he works for now. He came from a culture where older people's opinions were valued more than those of younger people. Nick's organization is a young startup. Thus, most of the employees they hire are straight out of college, young, and driven. Nonetheless, Nick operates under the assumption that seniority and decision-making power come with age and title, and that junior people are not supposed to be involved in strategic decisions. They are there to learn, grow, and prove themselves with concretely delegated tasks, given by the leader. As a result, his young team was never considered senior enough to be part of the discussion on a sales strategy. As this was not the common practice in such an organization, it created friction between Nick, his team, and individuals across the organization who could not understand his choice to hire an expensive consultant instead of brainstorming solutions with them like any other leader in the company. If Nick worked in an organization that valued individualistic leaders who only trust senior expert advice, no one would have expected to be asked in the first place.

When Nick understood the gap between the culture he came from and the company's culture, he understood the choice he had to make. Either he learned to collaborate with everyone equally regardless of their seniority level or he should leave and find a company that followed a more traditional hierarchy. Regardless of his preference, bias, or belief, his young team was not going to perform at their best if they were not included in strategic discussions. This would not only influence the way he was perceived as a leader in the organization, it would also shine on his team. The people on his team would be considered low performers because others tend to assume everyone behaves like them. Hence, it would be easier to assume that Nick involved his team first and they failed to than that he went directly for external advice.

When you understand your work culture, you can decide whether it is the right culture for you or not. As a leader, you can then become the advocate for this culture, supporting your team to make the same behavioral decisions for themselves. When people know what is expected of them, they can focus on the work itself rather than interpersonal conflicts.

Mission

What is the mission? What is the most important thing to achieve right now? What should we prioritize when everything seems urgent and important?

"If you have seized a lot, you have not seized" is a Hebrew Talmudic idiom to express the idea that when one tries to achieve everything, they achieve nothing. Having a purpose is inspiring your people for a better future, having a mission is focusing your people on the one step you all need to take right now.

Particularly during times of change, uncertainty, or a crisis. Prioritization becomes hard when organizations reinvent or change their business model, operating model, and culture. Everything is important and urgent. Moreover, importance can change frequently, sometimes on a daily basis. This is the key reason having a focus is so vital. A team without a clear mission will be swayed by any new compelling idea, start and not finish important work, and always need to seek the approval of external personnel, such as the leader's, to make decisions. Doing too many things is one of most common reason things do not get done.

You Are the Translator

As a leader, it is your responsibility to translate the overall company vision and mission statement to the day-to-day work of your team. You are not alone in it. When was the last time you asked your people to vote on the most important thing right now? Given a certain objective, what should we focus on? Your people know best what they are working on right now. They know best what blocks them. They know what they have on their to-do list. You have the destination and each one of them is holding part of the map. Together you can have a conversation on what needs to happen in order to reach a certain objective and goal. It might not always be straightforward, but it's best to have many brains on the same problem.

Since you are the leader, you are the one with the perspective. You know the destination or at least, you can define it or influence it. For you, it may be much easier to decide which path to take. However, you are too far to see the streets. You are not aware how many turns to take or if there is traffic on the highway. All you know is the general direction. Are we targeting large corporates or small ones? Do we prefer to focus on speed or finishing the work thoroughly? It is like a compass. You provide a direction, a focus that enables your people to finish things rather than starting new things because everything is always so urgent. When they understand the principles by which you make decisions, they will be able to prioritize what matters. Also, when priorities shift, your team can then confidently judge whether it is aligned with the overall company purpose and decide accordingly. A mission is only a focus towards a direction.

Do Not Assume. Clarify the Mission

Nick thought the mission was clear. It rarely is. He thought everyone knew that the company was aspiring to go public and everything they worked toward now was related to that. However, what does going public even mean? For some, it is a time to standardize and systematize. For others, it is a time to stabilize the revenue stream and company growth. For Nick's young team, it meant nothing. They had never taken a company public. Many of them were unaware of what it takes for a company to become public. For them, the day-to-day operation was the focus. The salespeople were there for selling; the product development team was there to build the product. Everyone had their to-do list. None of the items included ensuring the sustainability of the sales funnel.

What would they have done differently if they knew? For a company to go public, they must show a sustainable sales funnel. Hence, the company needed to focus on how to attract and retain target customers. Also, it should develop long-term relationships with them and learn from them how to improve the product so that they are less likely to lose any customers on the way. This is an ideal, not an expectation. You can't always know to that level of detail. However, if you do, share it. Never assume people know.

Case Study Epilogue

For Nick, the stages of emancipation were difficult. Through the process, he understood that he was unaware of the company vision, operated in a different culture than his peers, and never communicated the company business objectives to his team. His first reaction was to reject it and get annoyed that it was too idealistic. Ultimately, it meant that he didn't do his research before taking the job. Then he provided his team with the right information. Information that neither he nor his team knew they needed. It is much easier to hire an expert to fix the problem than invest days into understanding the vision, clarifying priorities, principles, desired behaviors, and learning to work collaboratively. It felt like too much.

Rome was not built in a day. Not all leaders do everything right from day one. Intention counts and taking one small step at a time is important. Even small awkward steps can get you moving in the right direction.

For Nick, it was to take his salespeople and product development team to an offsite. He felt awkward about it, but he did it anyway. The first step of emancipation is to accept that it is your job to provide people with transparent and honest information so that they, with you, can make decisions.

"We are here to solve a problem," Nick explained to his team. "We care about our customers; otherwise we wouldn't care about their problems. Caring is one of our company values." For the individuals present, it was the first time anyone explained the company values. "We are aggressive with our customers. We push the product to them instead of understanding what they are missing. Would you do that to someone you care about?" It resonated. Everyone cared about the customer. The salespeople felt uncomfortable pushing solutions that were insufficient, and the product development team was annoyed by all the urgent escalations by customers about features they never planned to develop. "Our goal is to find the middle ground. What are the minimum features we need to build now to not only increase our sales but also that our customers are happy to buy our product?"

The day was full of fruitful discussions. They discussed the product from the viewpoint of a customer. They argued about the problem the product solved and the type of relationship they were expected to develop with the customers. At times, it became uncomfortable. However, everyone appreciated being part of the discussion.

Throughout the offsite, the team asked many valuable questions that led to interesting outcomes. The role of a salesperson transformed into one of a success consultant who focused on how to help the customers to utilize the product better, rather than struggle on their own. Multiple individuals from the development team suggested being present at product demos to understand better where sales pitches failed. Also, they came up with a list of small improvements to address critical sales blockers. Nick gathered all of the questions and suggestions and was charged with positive hopes for the future.

As the sales improved, Nick decided to get a bit more vulnerable with his team. They went on another offsite to discuss their culture. The team appreciated it because they could contribute to new team norms, role definitions, and iterate on their mission, now that the sales funnel was more stable.

When leading a workshop to discuss the internal team operating model, it is always useful to think about the team as one part in a larger ecosystem. Remember the organizational purpose; this is where we want to be one day. Then remember the culture; this is how we want to behave, within the team and outside the team, with each unique stakeholder and customer. The mission is the last piece. What is the most important thing to focus on right now? Who is responsible for each part? Who is accountable and whose support should we seek?

Nick's team liked the change in him. Being naturally caring individuals, they appreciated his effort so much that they supported him to get better at collaborating. Whenever Nick wondered on his own, his team reminded him that they were there for support. "We don't need answers, just a direction" That enabled Nick to develop trust in his team, emancipating them to own many more strategic decisions going forward. It was not an easy journey; it took years and Nick still finds himself forgetting to ask for help. However, he knows it is not about being perfect; it is about slightly improving every day.

Conclusion

When a leader operates from a mindset of emancipation, they will not run outside for help. They will let their team decide when they require that support. People know best what they need.

Emancipation drives ownership. When you operate from that mindset, you will discover that the more you trust your people, the more they can be trusted. They develop a sense of ownership and can now focus on delivering better outcomes that are aligned with the company's vision and objectives. Also, the team as a whole will be perceived as a higher performing team by the organization, bringing glory to all.

For you, as a leader, emancipating your people gives you the trust that your team is working on the right thing and will be able to make the right decisions when given the responsibility. Nevertheless, emancipation requires more than the leader's willingness to let go. It requires the team to be able to pick up. Hence, the next step is to ensure your team's capability. You will learn all about it in the next chapter.

Questions to get you started:

1. What is your desired impact on the world?
2. What do you need to do to achieve your desired impact on the world?
3. Why do you want to create this impact on the world?
4. What does success look like in your current situation?
5. What would success look like in the future?
6. What behaviors do you value? What behaviors does your organization value?
7. What is considered normal behavior in your organization?
8. What behaviors do you aspire to have but don't necessarily have right now?
9. What is the most important thing right now?
10. What makes it important?
11. What would it look like when you achieve it?
12. How does your contribution or your team contribution support you or your organization in achieving your goals?

Ready for the next phase of the ACE model? We are going to dive into the concept of capability: flip to the next chapter and discover more.

CHAPTER 4

Capability

Challenge Your People to Grow

Capability refers to a person's power or ability to get something done or generate an outcome. Low capability suggests that a person may struggle to accomplish a particular goal or perform a particular task. High capability, on the contrary, means that accomplishing the goal or performing the task may come easier to that highly capable person.

Many people tend to confuse capability with skill. While a skill is purely based on one's ability to perform a task, capability is also their potential to develop that skill with the intention to generate an outcome or get something tangible done. For example, writing skill is one's ability to write well. Capability, however, is measured by the ability of that individual to do something with this skill such as write a paper or a book. Therefore, a capable writer will use their skill of writing to write a book. The quality of the book depends on their skill level in writing but not just.

High capability depends on more than skill. It builds upon one's potential to grow in that area, intrinsic motivation, and external support. For example, many people have the potential to become great writers, however, only few will write a book. Why? Developing this capability requires the fundamental skill to write, the intrinsic motivation to share your words with the world, and the support of others in seeing it through. A person's capability is typically a joint effort at life and work.

© Bar Schwartz 2020
B. Schwartz, *Leadership in a Time of Continuous Technological Change*,
https://doi.org/10.1007/978-1-4842-6300-6_4

Capabilities at Work

When was the last time you applied for a job? Or maybe you are hiring. Most job advertisements are structured in such a way where they first have an overview about the company or the team, then a list of activities the applicant will be part of, and lastly, a list of education and skills requirements. I call it the ideal shopping list. In practice, many employers rarely follow these lists because people rarely check all criteria. Also, it rarely matters. What matters is that you are capable of doing the work you were hired to do. Can you get things done?

I assume that you, just like me, have found yourself in situations where you were responsible for results you had no control over. Where you were incapable of achieving a concrete outcome. Maybe you had the skill to do it but lacked the influence, authority, buy-in, or support. Alternatively, you lacked the skill because the outcome required more than you could accomplish alone. At work, people often depend on others.

A person's capability at work goes beyond their skill set, education, or any professional background they bring. It is also not about setting up a process and coordinating all the individuals in it. It is about one's ability to know how to achieve a concrete outcome and being able, in practice, to do so. Hence, it can include developing a skill, changing a process, and bringing others on board to support it. Hence, understanding what someone is capable of doing and in what context is what matters for job performance.

Lack of Clarity at Work

You may have experienced a lack of clarity of roles and responsibilities at work. Maybe you worked in a fast-scaling startup, maybe you switched jobs often, or maybe your manager never had that conversation with you. Your actual job, the role and expected responsibilities in that role, may be different in practice than what was written in your contract or what you verbally agreed with your supervisor when you started. The bottom line is that when your role or expected responsibilities are unclear, your goals are also unclear. You don't know what is expected of you so you feel unable to deliver on those expectations.

Comparably, a platform that provides culture and compensation data for public and private companies, surveyed over 10,000 respondents in the tech industry to understand their biggest stressor at works. Unclear goals came in first with 42% of the respondents choosing it as their biggest stressor at work. It ranked much higher than having a bad manager (16%), long commute (16%), difficult co-workers (14%) and long hours (12%). Surprisingly, in the engineering department, the score was higher. In a subset analysis of 5,000 respondents from the engineering department, 47% answered that unclear goals are their biggest stressor.

Not knowing what is expected of you or having unrealistic expectations is a very common problem in engineering departments. Maybe it is because Engineering, IT and R&D departments are often the first ones to go through an organizational change to adopt a more agile way of working. After all, based on the 13th version of the State of Agile survey, 74% of the companies that implement any type of agile practices answered that their main reason to do so is to accelerate software delivery.

Organizational change typically influences people's roles, organizational structure, and work processes. However, not every organizational change happens systemically and is executed strategically from the top down. Most organizational changes happen due to the dynamic of the day-to-day operations. For example, scaling up the headcount is an implicit organizational change. By increasing headcount, you also create new teams. Based on how fast you scale, some people may shift from a team contributor role to a team lead role without sufficient onboarding time. Also, some teams will remain under-staffed because it takes time to hire the right person. This typically leads to a lack of certain skills. Hence, individuals with those skills will have to support multiple teams or be clustered on a separate team so they can work as a service team rather than an individual team contributor. See how that could easily become unclear?

As a leader, supporting your people throughout a change often means that you also support them to pivot themselves in the organization. This is the type of support I always wished my managers would have offered. This is the type of work people should not do themselves. When done wrong, people may feel as if they have no future in an organization. Also, it requires alignment with the organizational context, meaning people should be emancipated. There is a balance between the organizational needs and the individual needs that must be met in order to ensure that people remain motivated.

To understand the concept of capability in the ACE model, let's use Jeff and Lea as an example.

Case Study: The Story of Jeff and Lea

Jeff is an ambitious and enthusiastic engineering leader. Within five years, he has gone through multiple promotions from being an intern software engineer at a small startup to a team lead of one of the engineering teams at a large, multi-division corporation. It required switching organizations frequently, investing less than a year at every role. However, Jeff always lands on his feet. At every company he worked for, regardless of his seniority, team size, and the complexity of the product or service he was part of delivering, he has exceeded expectations.

Chapter 4 | Capability

When Jeff joined his new startup, A Corp., as the Head of Engineering, he was not surprised to see a very different setup than what he was used to. A Corp., similar to many other startups at their stage, scaled up their engineering department very quickly over the last year, tripling the headcount to compensate for lack of engineering capacity. However, scaling headcount does not directly scale capacity for execution. Their engineering teams were slower than ever. Furthermore, their results did not move the needle when it came to business outcomes. Also, the product quality was declining.

Lea is one of the engineers on the Development Operations team, known as DevOps. She is responsible for the product quality, primarily ensuring that nothing is released to the customer before it is tested appropriately.

For Lea, quality is more than a job. It's a mission. Her long tenure of experience in the field of quality assurance and testing has validated how important her role is in ensuring customers love the product. She knows it makes a difference and makes a point to educate others.

Lea's passion for numbers was not an expectation in her role at A Corp. Nevertheless, she enjoyed tracking multiple business and operational metrics. "The more issues we find late in the product delivery pipeline, the more calls we receive at our call center," she explained. "The customer service representatives really don't like us. There is a limit of what I can find in a day of testing. The rest, the customers find, and our representatives take the heat. If our engineers knew how frustrated our customers are, they would prioritize differently."

When Jeff met Lea, he was impressed by her attitude. He saw how Lea was working against A Corp.'s process and structure. She was the only engineer continuously reaching out to the call center to understand what customers were complaining about. Also, she was in daily arguments with the engineering team leads to prioritize and address certain issues.

The engineering team leads at A Corp. were often disconnected from the customer. Therefore, they underestimated the importance of the issues Lea raised. That frustrated Lea. She perceived it as lack of care.

It became concerning when Jeff started sensing that Lea might quit very soon. Her inability to influence the quality of the product and achieve better results was linked to her own feeling of success. Losing Lea would lead to many negative consequences for A Corp.

Deep Dive
What Was the Challenge?

Jeff wanted to help Lea build the capability to influence product quality. Achieving this in practice meant that Lea and the engineering teams had to have a shared, mutual understanding of how certain issues impacted the customer. This understanding would hopefully influence priorities so that the most important issues would get resolved faster.

Jeff felt unclear on the reason Lea struggled to get important issues prioritized. It was unclear what Lea did wrong when communicating with the engineering team leads. She had the numbers; she had concrete examples from the call center. Yet, the team didn't seem to listen and kept prioritizing other requirements instead.

Process Cannot Substitute for Capabilities

After long conversations with the engineering team leads and the customers service representatives from the call center, Jeff decided to bring them all into one room to discuss the situation. The mood in the room was disturbing. The call center team lead expressed his frustration with the engineering team leads, stressing out about how unable they were to build a working product. The engineering team leads responded with complains about how often the customer service representatives made "an elephant out of a mouse" by escalating issues to the top management. "It was just one pixel on the screen!" said one of the engineering team leads. "I want to see you working every day, all day, in front of a flickering pixel," replied the call center team lead.

The meeting was long. Lea tried to moderate it by presenting business numbers, trying to connect customer satisfaction, sales, and operational efficiency with the choices the engineering team made. At some point, Jeff suggested creating a new process where the team leads from both departments came together with Lea twice a month to discuss which issues should be prioritized. This would enable everyone to express their needs in person, while also provide a platform for Lea to influence the decision.

The process went well for a few months. However, as the time passed, more people were hired, teams changed, and it became harder to prioritize that meeting. The frequency became so rare that Lea found herself in the same situation as before, with a huge list of issues she could never prioritize.

Authority Cannot Substitute for Capabilities

Jeff's second attempt was to get Lea two dedicated engineers. He hoped that by resolving some of the issues in a focused way, the pressure on the teams would reduce. Also, Lea would be able to address some of the critical issues faster instead of waiting for the teams to prioritize them. She accepted the new setup with excitement.

A month later, the product quality did improve but only slightly. The call center reported receiving fewer calls but not significantly. They enjoyed working with Lea and her developers directly and the speed with which she was able to address their challenges.

Jeff was relieved to see the numbers. However, when he spoke to the developers, he realized that this solution was not scalable. While they were happy with Lea's leadership, they were unhappy with the scope of their work. Moreover, they felt like they were being punished by having to fix everyone's work. One of the engineers mentioned that it felt as if the teams stopped caring about their quality completely. Now that they had two dedicated people for cleanup, everyone expected them to fix all issues.

Case Study Conclusion

Local optimization might not always lead to overall optimization. Sometimes, incremental steps won't work. Jeff identified correctly that the challenge was enabling Lea to develop the capability to influence the product. Therefore, he tried to enable Lea by creating a process that supported her in raising awareness on the issues she found. When that became insufficient, he provided her with a small team to implement solutions for those issues. However, those issues were only part of a larger pipeline of requirements, requests, ideas, and issues that the engineering teams had to work on. With every new product release, the list of items to develop got larger and larger. Everything was urgent, everything was important, and everything influenced the bottom line. Every new release generated more issues on Lea's list. Moreover, more people, more teams, more issues. It became a never-ending chase after the desired quality.

So, what do you do when your people struggle to get things done? You focus on building capabilities. After all, enabling your people to grow in their capabilities is your priority as a leader. Results will follow.

Building Capability

What can you do to build capability or bridge the lack of capability with your existing people? The following steps describe

- How to identify the required capabilities
- How to create a sense of ownership to build capabilities
- How to maximize growth for individuals

Structure

Organizational structure influences how responsibilities are delegated, coordinated, and supervised. For example, functional teams normally are responsible only for their function while divisions or multi-functional teams are responsible for the whole product delivery since all the people necessary are in one place.

For leaders, understanding organizational structure is important since it allows you to determine what capabilities are required for each role. Expecting a designer to implement a software product alone is as unrealistic as expecting a waiter in a large restaurant to also cook every meal. However, expecting a team of engineers to also test their code is as realistic as expecting the head chef to wash dishes or cut onions during rush hour.

Analyzing Your Organizational Structure

Analyzing the structure of your organization, department, or team is a first important step in recognizing what capabilities are needed in each role. Often the structure, roles, and responsibilities are designed to support an organization-wide process to deliver a quality product or service in a timely manner. Hence, when your structure is fuzzy, analyzing the process is the alternative way to gather that information. Nevertheless, gather it. What needs to happen in your organization to deliver a business outcome, a product, or service a customer is willing to pay money for?

If your organization is large and you lead as part of a singular division, department, or team with no interaction with other parts of the organization, it should be enough to analyze only your core process and roles that are part of that process. At the same time, take the time to investigate the role other divisions, departments, or teams play in your product or service delivery process. You might not have any influence over them, but understanding this will allow you to prioritize what capabilities are more important.

When analyzing the structure or process, think deeply about your organization. Then, write down all the different steps and people or roles that are needed for the product or service to be delivered. Be as precise as you can be. If you do not know, ask.

Here are some questions to get you started:

- What are the steps an idea goes through from being conceptualized by person X to being implemented and released to customer Y?
- Which division, department, or team is part of which step?
- What is the deliverable of each step?
- Who is expected to do what?
- What department or team do they belong to?
- Who are they reporting to?
- What dependencies exist between teams and roles in a concrete step?
- What dependencies exist between teams and roles to move from one step to another?
- What is the key delivery for each team and role?
- How does communication happen between departments, teams, roles, and individuals?

For example, a simplified software product delivery process could include the following steps:

1. **Request submission**: Requests such as abstract ideas, specific customer requests, or internal stakeholders' requests, innovations, improvements, defects, and issues could arrive from anyone within or outside the organization.

2. **Evaluation**: Each request is evaluated by a group, a team, or individuals to determine whether we should or should not implement that request. In each organization, this step will look very different. Some organizations will have a steering committee; others will do it on a development team level. It depends on the number of people involved in the process. Also, the level of the evaluation differs. Some organizations will evaluate how aligned the request is with the business strategy, how it fits with the current state of the product, how desirable it is by a customer, how feasible it is to develop it in a reasonable timeline, and other criteria. Others may figure it out as they attempt to implement it.

Leadership in a Time of Continuous Technological Change

3. **Scoping**: When a request is being selected for implementation, it's often required to scope it or shape it before one can start implementing it. Many companies handle this stage within their development teams. However, depending on the level of abstraction, the team structure, and size, it might be an external step. Then the team may receive a very detailed requirement that specifies exactly what and how to implement the request.

4. **Implementation**: Assuming everything went well in the previous steps, an engineer will pick the request up and implement it. Based on the complexity of the request, the state of the existing product, and the infrastructure, the time it takes to build it right will differ. Often, new questions will arise during the implementation. If the engineer can reach out easily to the relevant stakeholders, they will be able to answer their questions faster. However, if they can't do this due to organizational structure, culture, or availability, they will either be blocked or make an intuitive guess. Moreover, this step may require collaborating with multiple people to complete one request. For example, one person will develop the customer-facing user interface, another will design it, and a third one will integrate it with the existing product.

5. **Testing and review**: When the implementation is complete, the final outcome is often transferred to a testing system or environment. There are cases where an engineer asks another engineer to test and review the outcome on their own local system or environment. There are also cases where engineers test their work on a live environment called production. However, it is recommended that the team has a testing system that simulates a customer environment in the best and most realistic way possible. Then multiple individuals take part in testing and reviewing. If defects, issues, or misunderstanding in the scoping are discovered, the outcome is likely to be rejected and the engineers are expected to fix it before releasing it to a customer. Sometimes, due to lack of time or if the issue is not too severe, the outcome will be released as is.

6. **Monitoring, maintenance, and updates**: This step varies the most. What happens after the software is available to a customer differs between business models. For example, when Facebook releases a change on their website, you get it immediately. All you need to do is open the website and it is there. If anything does not work, Facebook has tracking in place to identify where you are stuck. Also, customers can report issues on the website. However, products such as Microsoft Office or Dropbox have different delivery models. In Microsoft Office, one has to proactively update the software. In Dropbox, some features cost additional fees.

As you can observe, there is no way for me to capture this process concretely as it is in your organization without observing it in practice. There are so many ways this process can be executed. Depending on your organization size, complexity, number of product pipelines, business model, structure, culture, and many other factors, this process will be different in practice. Hence, to understand the role you and your team play in it, you start by gathering the data of where you are in that process and how your team contributes to the final outcome.

Product Development Capabilities at Jeff's Organization

Jeff struggled at first to clarify how Lea's team fit within the product development structure and process. It was a functional team in a system of cross-functional teams. For a tester to test a product increment on Lea's team, they needed two key things: an understanding of how to test the product increment and a working software version that included that increment in a testing environment.

To develop an understanding of how to test the product increment, the tester needed to work with the product manager and the designer to clarify a set of use cases and their expected behavior. They also needed a working software version from the engineer, who also needed to understand the same use cases and expected behavior.

The delivery or outcome of the tester was a green or red light to release the new version to the customers. In addition, a full report of what was tested and the actual behavior was used to keep track of that information in case any issue was reported later.

The stakeholders of the tester for this specific activity of testing a product increment were the product manager, designer, engineers, and everyone who received the report. It would be helpful if they had a strong relationship with the customer service team or the customers directly to ensure they could prioritize which issues were release blockers and recommend what to fix first.

To succeed as a tester, capabilities include skills such as communication in written and verbal form, traits such as curiosity to understand how things work, and knowledge of user's behavior and testing methodologies. However, if we think about testing as a capability rather than a person, then we understand that engineers are expected to have the same understanding of the use cases and expected behavior of every increment they deliver. Also, they have similar stakes since they work with the product managers, designers, and other engineers in their team. Hence, the core capability missing is the ability of an engineer to utilize testing methodologies and generate a test report. Think in capabilities, not in people.

Clarify Roles and How They Interconnect in a Workflow or Process

As a leader, it is vital for you to understand each role in the existing structure you operate in, including roles that do not report to you. Clarifying the roles of a company or a division is a joint effort involving representatives of each role brainstorming together. You are not alone in it. Often your human resources department will have an initial description of each role. Use that as a guiding document to make a list of capabilities you need to generate your business outcomes.

The outcome of a role clarification brainstorming session should at least clarify the following questions for each role:

- What is the purpose of this role?
- What makes this role important?
- What does success look like for a person in this role?
- What activities, tasks, or responsibility areas should a person in this role be responsible for?
- What activities, tasks, or responsibility areas should a person in this role be involved in?
- What decisions are owned by a person in this role?
- What deliverables is this person responsible for?
- Who are the key stakeholders for a person in this role?

After completion of the structure exercise described in this session and potentially the role exercise, you should have a visualization of all of the steps in your core process. For each step, you should have a list of the roles in that step. For each role, you should have the questions above answered the best as you can.

Start Small and Take Your Time

Acknowledge that it takes time. Gathering data, analyzing it, understanding it, and making sense of it takes time. If you lack the time, a quick win could be to only observe your team, similar to the way Jeff analyzed Lea's team, understanding the role of a tester in relation to the role of an engineer on the teams.

Look for similarities; shift from thinking about how an individual does a certain task to what needs to happen for that task to get done. You can connect people based on their dependencies, shared knowledge, or shared expected delivery. For example, as discussed, the testers have a strong dependency on the engineers in their team and the product managers, who are not part of the engineering department. They both must have the same initial knowledge and contribute to the same outcome.

You can also analyze one role. Understand the stakeholders around this role. What influences it, what enables it, and what information it needs to be successful. For example, what would a tester need to be successful? Then connect the dots from there.

Regardless of how you start to analyze your structure, remember to start small and give it time. Think about it as a marathon rather than a sprint.

The Value of Understanding Organizational Structures

Understanding your structure enables you, as a leader, to identify what type of leadership capabilities individuals can develop in their role. It also allows you to identify where you lack skill, knowledge, or staff. If you involve more people, looking at the problem together will generate better possibilities to address the challenge.

Understanding your structure also enables you to create growth opportunities for your people. When you know what capabilities are required for each role, you also know who in the organization should already have an advanced level of a specific capability. For example, Marketing would be a possible department to seek storytelling and communicating with clarity capabilities. Moving people temporarily to another team or working on a project together can be both an opportunity to practice a new capability and build relationships with people one doesn't often work with.

Setting up a project team with diverse capabilities and clear objectives for learning is a useful way to build new capabilities. Similar to the engineering teams Jeff leads, people from diverse engineering roles are expected in this setup to work together to deliver an outcome so they can practice activities and tasks that aren't a part of their role otherwise.

Leadership in a Time of Continuous Technological Change

To ensure success, it is important for each person to have the awareness of what they already do well and what they want to learn. Also, to ensure safety for exploring, the team should be evaluated as a team on the outcome they agreed upon and on their progress as a team in comparison to their evaluation from when they started.

Teams in such a setup are never comparable to other teams, even if they work on the same project or product. People are different and the level of capabilities is different.

How Did Jeff Analyze the Structure?

Jeff decided to address the testing challenge by moving the testers from the DevOps team to the engineering teams. He was aware that there were more teams than testers. However, by dedicating one tester to two to three teams, he encouraged the testers to act as trainers. Instead of testing the software on their own and generating a report, they focused on teaching the engineers on the teams to test each other's work. While some engineers refused to do it at first, over time they saw the value in it. Also, it encouraged them to clarify better what the product manager requested and stop accepting vague requests.

This step is the most challenging step in the ACE model. It also takes the longest to execute in practice. If you are up for it, you will be amazed how much you can optimize by shifting your mindset from what people are responsible for to clarifying their expected capabilities and enabling them, structurally, to get better in achieving their goals. Nevertheless, I do want to stress that you are not alone in it. Ask for support. Also, always start where you are, with the capabilities your team is responsible for before you start investigating the interfaces.

Jeff, for the sake of time, focused on the engineering department and the stakeholders who interfaced the closest with it. Also, he only specified the roles, responsibilities, and capabilities the engineering teams were responsible for. He explicitly excluded sales, marketing, customers support, and other functions that were part of the larger software delivery pipeline. His results were great for what he needed. Also, it gets easier to do overtime.

Clear Responsibilities, Clear Goals

As a leader, when the role or structure changes for individuals, make sure you communicate clearly to the person how their responsibilities will change as well. Sometimes it may be enough to clarify the new focus and expected outcome; however, more often, you will also need to go through the responsibilities of their previous role and compare them with the responsibilities of the new role. Make sure they know what is expected of them and can think with you about what is required to fulfil those expectations.

Remember: Clear goals. Unclear goals are the biggest stress driver for most people at work. Be clear on what you want them to do. Then ask them "what would you need to do this?" to highlight the gap. After all, capability is the ability to do something.

Knowledge

Knowledge stands for anything or anyone someone knows, including facts, information, and skills. One can know another person, a domain, or how to do a certain task. Often, knowledge is measured by someone's familiarity, awareness, and understanding of either practical or theoretical knowledge. High familiarity with a domain might suggest high knowledge in a domain.

Knowledge can be acquired in many ways, such as by education and experience. Structure, as discussed, usually influences what people know as their responsibilities and shapes the type of work they do and the people they interact with. When they move from one role to another, the knowledge they require also changes. However, a change of role does not equal a change of knowledge. Hence, now that Jeff moved all testers, including Lea, to the teams, he can't automatically assume that the testers know how to switch from testing themselves to teaching others how to test. Nor does it mean the engineers will know how to test right away.

Creating Learning Opportunities

As a leader, you want to create learning opportunities for your people; however, you are not responsible for their actual learning. Your role is to clarify with them what they should learn or know to succeed in their role, but if they don't follow up on it, it is on them, not you.

The primary question here is "what do you need to do this?" The answer can vary. The person may need support from you such as training, mentoring, coaching, and even introductions to people across the company. While it is common for leaders to get frustrated when they provide that support to individuals and teams but see no results, we must acknowledge that people learn in different ways. Therefore, they need to own their personal and professional development. They need to share with you what ways of learning are effective for them or not.

Clarify Knowledge Needs

The knowledge step in the ACE model is about clarifying the knowledge needs and setting up an accountability process that enables you to follow up without taking the responsibility for their learning. To do that, you have to put yourself in their shoes. Not literally, of course. Figuratively, imagine a situation

Leadership in a Time of Continuous Technological Change

where your role have changed. Your manager announces one day that you are either responsible now for another task or that you are moving to a different setup that affects your job. You are not sure what it means. You do not have experience being responsible for executing this new task or achieving this outcome. It is new to you.

Now imagine that your manager invites you to a meeting to discuss your development plan. Here are questions you could start with to inspire your thought process:

- What was your previous role?
- What were you responsible for?
- What were the core technical skills you needed to do your job?
- What did you need to know to be successful in it?
- What is the new role?
- What changed in your responsibilities between the previous role and the new role?
- What do you need to know now to be successful?
- Who do you need to build a working relationship with to get your job done effectively?
- What skills are important to perform your new responsibilities or tasks?
- What domain knowledge should you have?
- What organizational processes should you be familiar with?

Some of these questions you can answer yourself; others you will have to answer with your manager or people who work in a similar role to yours in your organization and other organizations. Some questions might not be relevant at all. Create a list of all the things you should know. Stop yourself from jumping to solutions. We will get there.

When you have the list, you can assess each of the items on your list. Then ask people who work with you to assess you as well. It is a similar process as a 360-feedback review where certain organizations ask your manager, peers, direct reports, and key stakeholders how it is working with you. However, this is a proactive feedback request. You do it once when you switch roles. The results are for you.

Getting evaluated by your peers is an important step. You can narrow down the list to the most important knowledge gaps or skills. Also, you do not have to ask many people. Start with two to three colleagues you trust. What is important here is to start getting into the habit of receiving this feedback and avoid relying only on your manager's perception of you or your own perception of you.

The last step is to choose the most important items where you received a low score. Then discuss your learning objectives with your manager and add it as a topic to follow up in your feedback talks. You can teach your people to do the same with you.

How Did Jeff Clarify Lea's Knowledge Needs?

The idea of working collaboratively on one multifunctional team was new to Lea. Also, the responsibility of teaching instead of executing the testing was new. Jeff supported Lea in writing down what she needed to achieve her new goals. For example, Lea wanted to understand better how the product manager worked. The main blocker to teaching engineers how to write better use cases and know what to test for was understanding what the product manager was asking. For her, it meant shadowing the product manager when they conceptualized and clarified the requirements before they presented them to the team. Also, Lea asked for online training where she could learn how to build a training program effectively.

Jeff urged Lea to ask the team for feedback. What did they think about her current teaching skills? Also, she talked to the product manager to understand the process of generating the requirements. The feedback from Lea's team suggested that Lea should take a product management course, not to learn more about how the product manager does their job, but to learn how to do it herself so she could educate the product manager on how to collaborate better with the team. Apparently, the team felt that Lea should not only teach them, but also the product manager. That highlighted the knowledge gap of the product manager.

Together, Jeff and Lea came up with a development plan and a roadmap for her learning. Every week, Lea followed up on what she learned. As she learned more, she also identified more opportunities for herself and the engineers on the teams to get better. Also, it became easier to integrate feedback into her journey and adapt.

Strengths

Strengths are the ability of a person to do something better than others, on average and consistently. We all have strengths. They are the things we do well naturally without noticing. The things we are frequently requested by others to help them with. When we are told by others that we do these things well, we think "Oh, this is easy. Can't you do it too?"

Many people are not aware of their strengths, have no language to describe them, and often mistake their weaknesses for strengths and vice versa. When we are in a role that enables us to utilize our strengths, we flourish. When we are in a role that is focused on our weaknesses, we feel drained. If we have no awareness of our strengths, finding a role that builds primary on our strengths is a difficult task.

Maximize Strength or Improve Weaknesses?

There are many books and research on the approaches of leading with your strengths vs. leading with your weaknesses. Some suggest we should focus on the talents we naturally have and develop them into strengths, while others suggest we should focus on our weaknesses so we challenge ourselves and grow as individuals. The ACE model doesn't take sides. Both approaches are valid given the objectives and goals a person has. The key message here for leaders is that knowing the strengths of the people you lead (and yourself) is important to maximize capability building initiatives.

When you know someone's strengths, you can place them in a role where they will naturally flourish or challenge them with tasks where they can work on the skills they need to succeed in their future roles. When someone is working only on things they are naturally strong at, they may get bored. On the other hand, when they are constantly over-challenged with things they are weak at, they may give up.

How Did Jeff Clarify Lea's Strengths?

Jeff took the time to discuss with Lea her experience. He then understood better what went well in previous companies and what did not go so well. He remained curious, asking her about the things she liked about her job and what she wanted to develop. Building on his knowledge from the Identity step of the ACE model, Jeff also provided Lea with some personality frameworks to support her self-discovery process.

In her previous role, Lea's strengths were identifying problems, clarifying what customers wanted, connecting with people personally, and explaining the value of fixing an issue with numbers. In her new role, she was challenged frequently as she taught the engineers and the product manager to do the same. Teaching was new to Lea. She invested time to learn how to do it well; however, it was not a skill she aspired to develop.

We can call it a gap or a weakness. For the current challenge Lea was facing, it was her weakness. In a different setup, it might not be as important. The question is how important it is for her to learn it?

As a leader, you want to ensure the people you lead (yourself included) have the choice to develop the capabilities required to succeed in their roles. However, you also want to ensure they grow capabilities, skills, and knowledge in areas they are naturally interested in or have the potential for.

Why? These capabilities will change and grow with them as their role changes in relation to their seniority level, organizational structure, and new responsibilities. Would you like to invest a lot of time and effort in developing something you don't care about?

It is true that the same capabilities that enabled you to succeed in one role might not be the same capabilities you need in your new role. At some point, you must ask yourself if this is still the role you want. The more you iterate on the first step in this chapter, Structure, the more opportunities you will discover to create opportunities for people to grow, change roles, take more responsibilities, and optimize the way they work. Keep experimenting.

Case Study Epilogue

Jeff had a long journey with Lea. Remember where they started. Lea was a test engineer on a functional team called DevOps, responsible for the quality of the product before it went live. She lacked the capability to influence the product quality because her role had so many dependencies on other people, including the product manager and engineers.

The first step was to bridge the communication gap. Jeff had the customer service team talk to the engineers, so Lea was not the only one with customer feedback in mind. That, unfortunately, turned out to be harder and harder as the company increased headcount.

Next, Jeff dedicated two engineers to work with Lea on critical issues she found. This solution contributed to the quality. However, the influence was minimal. Also, the engineers were frustrated because their work was to fix issues other engineers in the team created.

When Jeff went through the Structure exercise, he focused on Lea and her stakeholders only. He considered the final step of testing and reviewing. What should the preconditions be for such a step? Who should do what? and what dependencies did Lea have on other individuals? That led to a change in the role of the testers. They left their functional team and joined the cross-functional engineering teams. Also, their role changed from executing tests to teaching the engineers how to do so.

Supporting the role transition, Jeff and Lea discovered that it was not just about learning how to teach; Lea had to support the product manager better. She wanted to understand what that role entailed. Owning this knowledge gap led to Lea receiving feedback that she had another person to teach, the product manager.

Taking this last step, Jeff and Lea realized that Lea was not enjoying teaching. Nevertheless, learning how to support the product manager better inspired them both to transition Lea to a product manager role where her strengths with interpersonal relationships and love for numbers could be better utilized.

Today, Lea is the product manager of the team. She makes sure the team understands what they are building, the use cases, and the expected behaviors. She enjoys communicating with the business people to understand the larger picture. She can prioritize better. Lastly, she loves generating reports, including graphs and metrics that show, in numbers, how well the team is doing.

Conclusion

Capability is a person's ability to get something done, to accomplish or achieve an outcome. In the organizational context, capability is a joint effort because often a person unable to achieve the larger organizational results or outcomes by themselves.

For a person to develop a certain capability, they need to have the potential, intrinsic motivation, and support. While we could invest in everyone developing everything, it is not the best utilization of people at work. Therefore, we continuously aim to balance between the organizational needs and the person's needs.

Building capabilities in your organization and with your people requires you to go on an analysis journey to gather information on the structure, knowledge gap, and strength of individuals. This journey starts with Structure; Analysing the organizational needs by analyzing the organizational structure and core processes. This step compliments the Emancipation step where you clarified the organization's purpose, culture, and mission. Knowledge (what people know) and Strength (what people are naturally good at) compliment the Identity step in the ACE model. The difference is that you support your people to self-discover instead of supporting your personal self-discovery.

Building capabilities is a long and incremental process. To succeed in it, start small and iterate. With every small change in scope, process, or role, repeat it. It will get easier and faster as you get better in having these types of conversations. Also, knowing your people and their performance over time eases this process.

Questions to get you started:

1. What is your current structure?
2. What core processes are you part of?
3. What triggers the process?
4. What is the outcome of the process?
5. What steps are taken to complete the process?
6. Who is involved in each step?
7. What role do they have?
8. What makes this role important?
9. What decisions are owned by this person?
10. What do they need to do to complete the activity or task in this step?
11. What do they need to know to complete the activity or task in this step?
12. What knowledge and skills are required for this role?

Moving forward, autonomy is not an abstract concept; it is a vital aspect of the ACE model that enables leaders carry their people along through an astute decision-making process to be able to fully delegate work to them.

CHAPTER 5

Autonomy

Put Tomorrow's Leaders in Charge

Autonomy refers to the conditions of self-governing. Similar to Emancipation—the act of enabling others to make decisions without having to rely on you—the Autonomy step in the ACE model involves the willingness of others to take the authority to make decisions so they are able to work independently and fully own certain responsibilities. In other words, it's about them owning their work.

Trust is a prerequisite for autonomy, but most people tend to not trust others who are not emancipated, or they perceive as not capable. Why? Autonomy requires trusting the other person to make the right decisions because you have delegated the responsibility for the outcome but remain accountable for it. Hence, if you lack trust in someone's capability to do the work, when you delegate to them, you probably won't delegate effectively and fully because you might feel like you have to be informed of every small step. You might struggle to let go of being in control of the situation. Also, if the person is not emancipated, they might not be able to make the right decision. Thus, they may end up validating your belief that you should not trust them. Often, this is the moment when you feel the lack of alignment and the blaming game starts.

When an organization is still in its initial start-up phase, autonomy might not be a consideration because leaders at that stage are required to be very hands-on. Also, the time it takes to get input of someone from leadership is much shorter because alignment and communication on smaller teams are much easier than in a larger corporation. People may feel that they have autonomy merely due to their close collaboration with the hands-on leaders.

© Bar Schwartz 2020
B. Schwartz, *Leadership in a Time of Continuous Technological Change*,
https://doi.org/10.1007/978-1-4842-6300-6_5

As the organization grows, the need for people and teams to work autonomously becomes more relevant and critical. Leaders start shifting into more operational and strategic roles, and have less time to collaborate effectively with every team member. At the same time, when one maintains certain responsibilities for years, it is hard to let go of those responsibilities and delegate them.

On the other hand, people get used to not having to take ownership due to so many leaders who struggle to give ownership. Thus, when they suddenly have to, they are either not ready or not willing. The default reaction of many leaders is then to "save them" from this struggle by just not delegating. It creates a reinforcing loop of leaders taking the full control, ownership, and responsibility for everything, leaving their people lacking autonomy.

As Brian Tracy wrote in his book *No Excuses!*, "Bad habits are easy to form, but hard to live with. Good habits are hard to form, but easy to live with." And as Goethe said, "Everything is hard before it's easy." People often make the choice to keep behaving or working in a certain way that they know is inefficient because when they think about changing it, it feels too hard, impossible, or frightening. Alternatively, they throw people into the deep water and say "Congrats! You are responsible for this now. Go figure how to do your job. Bye!" This is usually another recipe for disappointment.

Creating autonomy is a process that starts on the first day someone joins a company and continues throughout their entire time with the company. It requires leaders to slowly transfer responsibilities to their people while the person is required to slowly accept those new responsibilities and fully own them. This is how people align themselves with their new job. However, it cannot be forced. Hence, if your people do not seek to gain autonomy at all, stop right here and start investigating whether they feel emancipated and capable. Ideally, when someone understands the purpose of what they do, is aligned with the culture, cares about the mission, and feels that they are in a role that they are capable of doing, they will also seek autonomy.

To understand the concept of autonomy in the ACE model, let's use Adam as an example.

Case Study: Adam's Story

Adam is the CTO of a growing company in the software industry. He co-founded his company over 10 years ago; it has grown from a university project to become the leading company in its sector. Over the years, they have hired hundreds of people for their headquarters and worldwide. In recent years, the focus has been on hiring experts fast to both compensate for the high number of juniors at the company and to bring new knowledge and expertise to accelerate growth. At some point, it became a routine to onboard over 10 people per month, doubling or tripling the company size in a year.

Leadership in a Time of Continuous Technological Change

Growing is never easy. While Adam feels proud of everything they have accomplished as a business, he remembers fondly the times when the whole company could fit into one living room. The times when everyone knew everything and everyone. The times when people felt included and enabled to easily generate meaningful impact. He misses that.

Growing also comes with a cost. As the headcount increased, changes began to take place sporadically within the organization, causing Adam to feel anxious every day. For example, customer expectations of the product changed, so new and complex requirements started to arrive at an unrealistic speed. Also, people in different departments started to silo themselves, shifting slowly to an us-versus-them mentality, blaming each other for failures. Mostly, blaming Adam's department.

Adam was not immediately aware of these changes. The expectations of him started to become explicit only when the company crossed the 100-person mark. Because Adam's department had the highest headcount, the CEO and CFO asked him to measure his department performance to better explain how each new person in the technology department was being utilized.

As one of the founders and the CTO, Adam was deeply involved in the product delivery process from the very beginning. When the team was small, he was the first developer. Every person who joined the team was interviewed by him. Also, he was very hands-on in reviewing most of the product increments. Hence, the performance of individuals was measured based on their ability to meet deadlines and their ability to get along with Adam.

When the team grew and started splitting into multiple teams across multiple products, it became harder and harder for Adam to be involved in every conversation, review each product increment, and have a personal touch with each new developer who joined the team. They hired many new people in engineering to accommodate requests from larger clients and deal size. They hired experts with capabilities that existing people on the teams were not familiar with. At some point, Adam started questioning whether being so hands-on still made sense. The business needed him to become more strategic. However, with so many tactical responsibilities on his hands, he struggled to let go of being involved in everything.

People management was one of the responsibilities Adam struggled to delegate the most. Being the CTO, the whole engineering department reported to him. Clearly, that was much easier to handle when his team was small. Also, not much management was needed when everyone was aligned, engaged, and enabled. Now that the team had grown tremendously, Adam was barely able to have personal conversations with each individual on a yearly basis. Unfortunately, it meant most of his employees felt uncared for.

Complaints started to arise. However, Adam had no capacity to support every individual with their role clarification, career conversations, day-to-day challenges, and interpersonal conflicts. To solve that, he promoted his most trusted developers to take over a team lead role, delegating to them the daily responsibility for their people. The idea was that a team lead would maintain personal relationships with each person on their team and unblock day-to-day, team-level problems.

On paper, the solution made sense. It was a natural evolvement to add a new level of hierarchy to ensure each person had a manager to care for them. However, there was one main problem with this solution. None of the developers had experience in such a role. Therefore, Adam chose to not trust them with managerial responsibilities and authority. Decisions such as hiring, firing, and promoting individuals remained in Adam's hands.

How did it look in practice? The team leads were responsible for conducting individual conversations with each person on their team, approving vacation and sick leave, and providing performance-related feedback to Adam. The employee talks, where the feedback was discussed, would be conducted by Adam every six months. During those conversations, people had to convince Adam that they did well, and Adam would decide whether to promote the person, keep things as is, or let them go.

Some of the new leads liked this new challenge, while others would have preferred remaining in their technical role. As all of them had a long tenure with the company and close friendship with Adam, they agreed to this arrangement. They hoped it would work out; however, they all had their concerns. It was their first leadership role. Their only preparation was one weeklong leadership training course.

At first, the solution was sufficient because the team leads could easily escalate important topics to Adam directly when they were not able to make decisions. However, when the number of teams grew, Adam started investing even more time in aligning and coordinating work, addressing customers' problems, interviewing potential candidates, hiring, and resolving critical performance issues and interpersonal conflicts, leaving no time to support most of the team leads.

Together with the headcount growth, the business complexity and product also evolved. The company started to hire new roles and more diverse individuals. As a result, people's expectations evolved as well. New requests emerged. People wanted to have role clarity, a career path, training, personal and professional development, and advanced titles that reflected their performance. While Adam was familiar with software development roles, he lacked experience with the supporting roles such as quality assurance, product management, design and agile coaching.

Despite his best intentions to support everyone, many people in Adam's department felt neglected. Rumors about nepotism started spreading. That led to some great talents leaving the company. Also, it led to "free riders," people who were doing the bare minimum because they thought no one cared about their work and that no one would notice. Lastly, it created confusion about people's capability and performance. Many great people felt unable to do their job either due to not knowing what their job was or due to not having any authority to make decisions.

The situation was not the same in every department at Adam's organization. Some leaders thought they were doing a better job than Adam. Others were frustrated by the lack of organizational clarity, slow responsiveness, and perceived lack of discipline of certain individuals. Thus, they openly criticized his capabilities as a CTO and the quality of the people in the engineering department. People were called lazy, incompetent, and weak. That created a psychological gap that influenced communication and collaboration across departments.

To understand better what led to this situation, let's take a step back and deep dive into how the concepts of autonomy, control, trust, and delegation interconnect.

Deep Dive
The Connection Between Trust and Delegation

So why did Adam keep the authority for so many important decisions for himself? Why didn't he train his team leads or hire more experienced team leads? The simple answer is lack of trust. Whether he was aware of it or not, trust led Adam to select inexperienced leaders for these roles. He trusted those individuals. Lack of trust was his reason to not delegate fully.

Yes, it is possible to trust and not trust people at the same time. We can trust people with specific tasks but not with others. Adam could trust those individuals to care for the people, manage the work, and provide performance feedback. However, he did not trust them with financial decisions such as hiring, firing, promoting, and negotiating compensation. His awareness and reasons varied from person to person.

There is a leader like Adam in most organizations. Leaders who struggle with trust. They can lead a team, a project, a department, or the whole organization. In a way, we are all a bit like Adam. Most people struggle with trust. Some more than others. Some are also less aware of this than others.

Many people fear delegation. Not only people who struggle with trust. The reasons for this fear vary. Some people can justify it as "people are not ready" or "need more hand-holding." Alternatively, they can argue that they are "faster doing it themselves" or that the "quality would be better."

Whether you want to be in control of the outcome, you want to protect others, or you want to save time, what it means in practice is that you do everything. Maybe you have good intentions. However, as a famous proverb says, "the road to hell is paved with good intentions."

When leaders struggle to delegate effectively, people feel it as lack of autonomy in their job. When people feel a lack of autonomy in their job, they often associate it with lack of trust. No matter how much you do trust.

Inexperienced or not, the team leads on Adam's team were more frustrated by their inability to take actions than their lack of experience in doing the role. They wanted to make hiring decisions. They wanted to be able to retain and acknowledge their best performers and address those who misbehaved. Waiting on Adam created bottlenecks and resulted in certain issues escalating for no reason but the time it took to address them.

When leaders struggle to delegate effectively, it can cause damage to the organization that is not apparent at first and is extremely hard to fix when becomes apparent. At first, it can appear as general frustration, speed decline, and top talents leaving the company. As it evolves, the remaining teammates over time lose their sense of ownership, getting used to not having authority and impact. Lastly, also the leader pays the price. Many leaders who struggle to delegate can't be sick or take vacations, suffer from anxiety and stress, and even experience a burnout.

If you feel that your team can't work if you are not around, be honest with yourself: how much do you trust your team? Is your trust level appropriate? Let me challenge you that you can always learn to trust more.

The Connection Between Control, Confidence, and Autonomy

What is in your control? When I explained the concept of emancipation, I referred to the process of children growing up and getting emancipated from their parents when they reach adulthood. I wrote about it very briefly, so you get the idea. Here, I want to dive slightly deeper into parenthood.

Kids are not born with the same capabilities as adults. This is clear. When you are born, you need your parents for everything. They are your main source of care in this world because they provide you with essentials like shelter, food, and attention. Also, they are your only source of information about the world in your early days. This is the reason kids tend to think of their parents as superheroes.

Leadership in a Time of Continuous Technological Change

Growing up is a process. You learn to walk, speak, brush your teeth, use the toilet, eat by yourself, and engage in early social interactions with other human beings. As you mature, you start behaving according to your personality. Some of it is a given; it is in your genes. This is often referred to in psychology as your nature. Other traits are learned through your culture and direct environment. This is referred to your nurture. I touched on this briefly in the Emancipation chapter.

What is in your control when you are a kid? Not much. You do not control what you eat, where you sleep, and sometimes you are not in control of the people around you, including the other kids you are friends with. It is decided for you by your parents' place of living. Also, kids often become friends with the kids of their parent's friends. Hence, not much is in your control.

When you mature, you start making your own choices. Some parents teach their kids to do that early on; others wait until their kids rebel. Nevertheless, this process, when you start gaining control over your own life choices, is the emancipation from your parents' process. You are now in control of your life.

What would have happened if your parents were not there to teach you how to dress yourself or if they never let you choose what you wanted to eat? Kids with over-controlling parents tend to be afraid to make their own choices. Alternatively, they rebel but then they make the choice for the sake of the choice, not because they know what they want. Kids with over-controlling parents tend to lack confidence in themselves and their ability to make decisions.

The same happens with employees. People often start their job as a junior. They learn their profession in university, a course, or on the job. They join a company or become self-employed. No one starts their career fully able to do the work they were hired to do. Everyone learns it somehow. Depending on the type of job and the person, some learn faster than others.

As a junior, what is in your control? Not much. Your boss will often be more experienced than you. Someone on your team or organization will teach you what to do. You might get very direct, specific, and precise tasks to work on. If your boss is hands-off, you might have to figure it all by yourself. Nevertheless, unless you have no boss, often you are not expected to make many decisions as a junior. Your priority should be learning.

What tends to happen in organizations is over-parenting. People are not junior forever. They learn, they get specialized in their job, and they are perfectly capable of making choices. However, the fear of some managers leads to many people never getting the opportunities or authority to make decisions. Their work is not in their control.

Case Study Conclusion

When your work is not in your control and you do not get to make decisions, you may start losing confidence in your ability to make these decisions or you may move on to a place where you get more control. Autonomy, ultimately, is having control over your work. Confidence is vital to ensure people can utilize being in control. Let's discuss further how you can create autonomy using control and confidence.

Creating Autonomy

The ACE model is designed to support leaders to effectively align and enable their people to realize their organizational vision. Creating autonomy, as the last step of the model, assumes that everyone is emancipated and capable. Now, it is all about how to intentionally inspire and enable your people to operate autonomously, assuming they want to.

Control

Control refers to the power to influence priorities or the authority to direct people to work on certain tasks in certain ways. The person in control is the final decision-maker. Therefore, they should also be the ones who understand the risks they can or can't take if the intended outcome is not achieved. Also, they should understand what success looks like, what is important, and what is nice to have.

As a leader, you are the owner of your team's outcome. Everyone on the team has the responsibility to contribute to the outcome but you are accountable for it. You take the risk by default.

For some people, control is a hard thing to give up. Certain people are more risk-averse than others. Also, the more mission-critical a task is, the more risk it entails.

Everyone Is In Control

Let's take an example. Imagine you are back in school. Your teacher gives you a group assignment to make a presentation about a topic you are unfamiliar with. Understandably, your classmates are also unfamiliar with the topic. Thus, you all have to invest time to learning it. To save time, you split the work between all of you. Each person gets a few slides of the presentation to make. The night before the submission, you realize that one of your classmates missed an important slide. Not including this slide will cost you 20% of your grade. You could do that, or you could work through the night to finish it. You hope the rest of the work done by your group is enough to pass in general. What would you do?

If you are risk-averse and you care about your grade, you will probably work through the night. It was a mistake of your classmate, but the grade is also yours. Losing 20% just like that? No way. However, what if you lose 49%? Would you do the same? What if you lose only 5%? What is the smallest grade loss you would be willing to risk?

What about your classmate? You trusted them to complete their part of the work and they missed this slide. Would you see it as an honest mistake? Would you see the person as reckless? Would you trust them again if you were to have a new assignment together?

This was a school assignment. Everyone shared the grade and were responsible equally for the final outcome. However, how often is that the case in organizations? Let's explore a different but similar example.

You Are In Control

Imagine you are an engagement manager at a consulting firm. You are staffed on a client engagement with a very challenging client, in an industry you are not familiar with and in a project scope you never did before. On the project, there are two juniors and two experts. For the juniors, it is their first project. For the experts, it is the first project with your firm. This is a very important client for the firm. Your performance was so great in other engagements that the partner thought it would be a great opportunity for you to shine and show off your skills and thus demonstrate that you are ready to be promoted.

You need to kick off the project. The team expects you to explain the project scope, timeline, and budget to them. A large part of the kick-off also includes defining who is doing what on the project. Everyone wants to do well but remember, you could get promoted. How would you split the work? How often would you check in with the team? Would you feel comfortable with the junior person working on their part until the last day? Would you feel comfortable with the expert doing that? If you are risk-averse and want to be promoted, I can guess that you will not. I can guess that you will want to review everyone's work as frequently as possible to ensure the quality is being met.

Would you behave differently if you were the expert on the team instead? As an expert, you are responsible for your part of the work. While everyone is there for the project as a whole, you are likely to still focus on your part more than you would focus on other's people's work. Assuming each of you could work completely independently from each other, how frequently would you check in with the whole team? What if it only matters how well you do your part because you are not evaluated in any way on the final outcome?

Who Should Be In Control?

There are no right or wrong answers here. Control is a tricky thing. In principle, we want to be in control if we care about the outcome and we are the ones taking the risk or getting the reward for the outcome failure or success. However, we might not be able to have this control.

In the school project situation, everyone had equal control because everyone got the same grade. In the consulting situation, the engagement manager had control of the project as a whole and the allocation of work between people. The expert would have the control for their part of the work only if the engagement manager allocated it. Hence, there can be a situation where the expert would not be in control of an outcome they cared about.

As a leader, you often have the control, power, and authority in your area. Thus, if you want people on your team to have autonomy, the way to achieve autonomy requires first deciding what should be in the power of others and what should be in your power. There are cases where you have several people reporting to you who can take the same responsibilities, so you need to analyze the right allocation of responsibilities.

Prepare a list of all of your responsibilities and ask yourself the following questions about each one of them:

1. Who should make the final decision?
2. What makes that person the right one to make that decision?
3. What capabilities are required to make that decision?
4. Who has the capabilities to take that decision?
5. What is the impact of me taking that decision on the team?

Note that certain responsibilities can be shared, and the decision can be made as a team. Then it is recommended to decide whether a team vote is required or a full consensus.

Again, the right person to make a decision is the person who cares about the specific outcome and takes the risk for it. Often, the person that is evaluated on it or is doing the job is the most suitable in the long-term since they also understand the decision better than someone who cares and takes the risk but is not familiar with the work.

What If the Right Person Can't Be in Control?

Thinking about Adam's situation, Adam is no longer part of the development team in engineering. He's not working with the engineers on projects and has no visibility on how well people perform their work, particularly when the people in the teams are experts in roles Adam has no experience with.

Adam cares but can he really understand the risks? Owning the employee talks with all of the people in engineering, Adam gathered information every six months from the team leads and sat with each employee to discuss their performance. He was the only one to approve promotions, changes in roles, and decisions to let someone go.

What makes Adam the right person to have the employee talks? Despite being the CTO, in the current company size, he met most of the people once during their hiring process. The feedback he provided was based fully on the team lead's assessment. However, the team leads were not part of the meeting. Also, as he could not follow up with every individual, the team leads were the ones to carry the follow-up responsibility after the talk. Lastly, individuals who were not familiar with Adam were nervous to receive feedback from the CTO since they didn't have the same informal relationship with him that others had.

As the CTO, Adam does take the risk of promoting the wrong person, giving an incorrect performance evaluation, and letting the wrong person go. He is accountable for the whole engineering department. However, is he the only person who cares about the failure or success of those activities? No. The team leads care more. They have a direct relationship with each individual. They also paid the highest price if someone left, became demotivated or was not held accountable for low performance.

Adam asked himself the following questions:

1. Who should make the final decision to conduct employee talks? Probably not him. It was not scalable.

2. What makes that person the right one to make that decision? A people skillset was needed to succeed in this type of role, skills Adam was still learning.

3. What capabilities are required to make that decision? Once the primary capability was to continuously partner with each individual to develop in their role, including understanding what they are doing now, what they would like to be doing next, and creating a measurable development plan for them.

4. Who has the capabilities to make that decision? In the current organization, the team's leads are slightly more capable of supporting individuals because they are closer to them and more aware of their work.

5. What is the impact of me taking that decision on the team? Adam was too far away. He also had too many people reporting to him. He could not partner with them frequently enough. Also, he lacked the time to invest in getting to know everyone. Most decisions were done based on gut feeling and what other people said about that individual.

Realizing that he was not the right person to conduct the employee talks, Adam decided to give away the control. It should be owned by the team leads. The only reason he kept owning it was his fear of the readiness of his team leads to make decisions involving change in compensation. However, ultimately, they were the right people to have the control.

Remember that capability is a foundation for autonomy? In a lack of clarity, Adam is the type of a leader who prefers to do nothing rather than make a mistake. Thus, he overlooked the options of building the capability of his team leads or hiring new leaders who could bring that capability with them.

When the right person to make the decision does not have the capabilities to do so, you should build the capabilities or place someone capable into the role rather than keep the decision for yourself. You should not delegate control to someone who is not capable or interested in having the control. Nevertheless, the right person is still the right person.

Remember, autonomy doesn't breed irresponsibility. You are still responsible for the outcome. You still have the power to overrule someone when you feel the risk they are taking is too high. You can demand an overview of the process and be involved or informed.

Control can be given or taken. Make sure to place people with the right capabilities in the roles you delegate responsibilities for, people whom you trust and want to develop further.

Capability and trust are fundamental for effective delegation. If both don't exist, ask yourself whether this is the right person for this role or would they be happier doing something else.

Giving Control

When you delegate, the other person takes over the active responsibility and you become an accountability partner. If the person is new to the role and responsibility, you can start sharing the control by setting up continuous check-ins where you coach them through the decision-making process, get updates on their progress, and can intervene when the risk is too high for them to grasp.

Keep in mind that they will make mistakes. Those mistakes are important learning lessons to recap on during your check-ins. Coaching people to reflect on mistakes enables them to create new learning for themselves instead of repeating the same failures. That can only be done when they are in control.

When the person is more experienced than you in making those types of decisions, it's recommended to use those check-ins to learn from them. Humility is acknowledging that we can always learn something new. Leading someone more experienced than you requires humility and courage to be open, explorative, and not perceive them as a threat to your role or power. Let them coach you and stand behind their decision, even when it's not fully aligned with your approach. You hired them for a reason.

To summarize, the first step to creating autonomy is to effectively assess what is in your control. Then, ask yourself the questions of what should be in your control and what could be in the control of others. The main principle to give someone control is to ensure that they care about it and understand the risk they take. Don't delegate control if you feel that a person is not trustworthy or capable, because you remain accountable for the outcome.

When you delegate control, you don't lose it, you extend it. It is all about trusting your people, enabling them to grow and have ownership of their work. So, lead with trust and allow others to be in control. Just ensure they are ready.

Confidence

Let's say you challenged yourself and trusted someone who turned out to be not trustworthy. Maybe your overall confidence in your people is low because they seem incapable and unwilling. Maybe you went the extra mile and gave them the control. Instead of taking it, they refused it or got overwhelmed by it.

Maybe they are not ready. Maybe they are not as capable as initially perceived. Responsibilities can only be taken, not given. The obvious choice would be to take it away, wouldn't it?

Confidence stands for the feeling or belief that one can rely on someone else to do something. At the ACE model, confidence refers to both the confidence of the leader in their people and the confidence of the individual in themselves.

Giving control is one step in building confidence as you shift power to the people you trust, enabling them to practice being in control and gain confidence over time as they deliver positive results.

When You Lack the Confidence to Give Someone Else Control

Lacking confidence in the control you give can lead to worse outcomes than not giving the control initially. One outcome could be that the person will feel the lack of trust in them. That can lead to lack of transparency, reduced motivation, and general frustration. Some people will take it personally. Others may see it as a sign that they are not really in control.

Adam decided to hire new people to take over the responsibility of people management and leadership. It was a long and hard decision to make as he was concerned that the existing team leads would see it negatively. While he gave the option to all existing team leads to get training and build their capabilities in the area, he was positively surprised when most of them chose to take a technical role without people management responsibilities.

It took a few months to find the first new team lead. Initially, Adam was not sure what he was looking for. The position description changed frequently. However, when the right person for the job appeared, the decision was clear. Everyone was happy to welcome Fabian to the team.

Fabian could not cover all teams. Thus, the existing team leads remained in their role until a new lead was hired. During this time, Adam and Fabian agreed on how to start shifting control from Adam to the team leads. The first step was the final decision on hiring.

Understanding the importance of delegating control, Adam decided to take a leap of faith by delegating this decision to Fabian and one other team lead. He felt that the team lead was capable of making such a decision and it was time to give the control.

A few days later, a new candidate passed all interview rounds. Most of the team liked him and wanted to hire the candidate, except Adam. When everyone came together to discuss the decision, Adam made a very elaborated case on how this candidate wouldn't be able to teach him anything. The team lead was frustrated and agreed to not hire the person. In the next interview round, the team lead waited on his decision until he received feedback from Adam, not making the decision himself. Ultimately, he wanted no conflict with Adam. Adam was puzzled.

Fabian's experience was different. When another new candidate passed all interview rounds and most of the team liked the candidate except Fabian, Adam did not object to the final decision to not hire the candidate. The team lead was frustrated. He could not understand what made Adam trust Fabian more than him.

Delegate with Confidence and Stick to It

Autonomy requires both sides, you and the other person, to accept the change in control and have a sense that they will figure out how to handle the newly delegated responsibility. If one of you lacks confidence in the other person, every small mistake or fear might trigger shifting back to the previous state. It is a paradox that many people must have control in order to start making decisions to develop the confidence while lack of confidence may trigger losing the control.

More so, the way you communicate the delegated responsibilities matters. Some leaders speak with teammates, staff, or subordinates as though they are not capable, but the leader has no choice but to delegate. If you don't trust your people, don't delegate. Start by addressing the reasons for the lack of trust. What are the assumptions that led you to the mistrust? What would need to happen for you to trust again? Speak with your people openly and listen to them carefully. Sometimes our lack of trust in others tells us more about ourselves than about the other person.

Messed it up? Adam delegated the hiring decision to the team lead, yet the team lead was not the one to decide. When the situation became clear to Adam, he apologized. He ensured the team lead that he trusted his decision and that even if he objected to a candidate, the team lead was the one to make the final call.

Confidence goes both ways and is built over time. How you delegate is as important as the delegation itself. When you delegate to your people with confidence, you are assuring them that you believe in what they can do, and this communication pattern itself can ignite faith in your people, causing them to go beyond your expectations.

Giving someone control means that you are confident the outcome they will deliver will be sufficient. They might make mistakes. These mistakes are great learning opportunities to build their capabilities further. Nevertheless, who wouldn't become confident over time if they know their leader has confidence in them?

Verifying

After asserting control and instilling confidence, verifying is the next foundational step when creating autonomy. Verifying refers to being explicit, precise, and coherent, ensuring you repeat the context and clarify the objectives and expectations. Verifying reduces the risk of misunderstandings and increases the possibility of successful delivery of delegated responsibilities.

When people ask for further clarification, explanation, or understanding, some leaders confuse it as a need for "hand-holding" or reducing the scope of the work. This confusion is understandable if you are under the impression you are being clear. Thus, the only reason the other person seeks more context, information, or further explanation is that they lack the understanding needed to complete the task. While sometimes it is the case, more often the lack of shared, mutual understanding comes from differences in communication styles. Hence, if you are not understood, all it says is that you need to invest more time into understanding the other person and explaining yourself in their language.

When I wrote about clarity in Chapter 1, I explained that clarity is the foundation of alignment. Clarity is measured by how clear something is to the other person, not to you. Therefore, verifying also means clarifying that the other person understood you. If you verify that they understood you and they have all the information they need, you can trust that they can work autonomously.

There are three parts to verifying. Being explicit, precise, and coherent. Let's verify that you understand all three.

Being explicit means that you say directly what you want. Try to avoid vague sentences because you think that they provide the other person more freedom to explore. Also, do not assume people understand things from the context. Do not rely on common sense, particularly if your working relationship is new. If something is not open for discussion, it should be told, not assumed. Things are never implied.

Being precise means that you use exact language to accurately express what you want. Leave little room for ambiguity. There is a difference between saying "please explore solutions for this" and "please explore what we can implement in two weeks to address this specific problem the customer is having." The focus is on something to implement in a clear timeframe for a concrete customer who has a concrete problem. We will discuss this more later.

Being coherent means that you also share your reasoning. What are your intentions? What is your context? Why do you need to delegate this to start with? Often managers and leaders try to test their people. Have you ever delegated something to see how the person would approach it? Would they go the extra mile? Unless the person knows that they are being tested, my opinion is that testing employees should remain in a job interview and training context, not when you are working towards a concrete outcome. The risk of that expands beyond the possibility that you won't get what you want. It also reduces the other person's confidence and trust in you as the leader and keeps them constantly alert and stressed.

The three aspects of verifying are the three famous questions: why, what, and how.

Why and What Makes Something Important?

When validating whether something is clear, start by asking why something is important, then follow it by asking what makes it important. *Why* reveals the context and focuses the discussion on the past information that is relevant to understanding the importance of a successful delivery. At the same time, understanding *what* makes this the most important thing to do right now is important. *Why* connects you to your purpose, *what* connects you to the current mission. Hence, verify both the why and the what.

Many leaders assume that since they said something during a meeting a while ago, people should remember. Even if you just went through the ACE model and clarified the purpose, mission, and culture for your team, many people will forget. The day-to-day work takes over and it is easy to get distracted by urgent fires and shifting priorities, especially when your environment is continuously changing due to the reality of your market and ongoing transformation initiatives.

As a leader, you are the one responsible for reminding your people of the context every time that you delegate something. Link the current thing you are working on to the larger purpose and mission of the organization. Explain the reasons that led to certain decisions. Your people need to know why they should work on this and what makes it important to you. If you can also explain your reason to delegate this task to this specific person, you get extra points. Of course, the longer you work with someone, the less you will have to explicitly say. However, I am a believer that too much information is better than too little information.

When Adam delegated the hiring decision to the team lead as part of his desire to give autonomy to the lead, he gave him the control. However, when the decision was on the table, Adam's confidence disappeared, and he became anxious and fearful that the team lead would not make the right decision.

The team discussed the candidate for a brief 15 minutes. Most individuals evaluated the technical expertise of the candidate without considering the candidate's behavior, perceived way of thinking, and values. The team lead, as a tenured employee, shared that approach. That led to Adam to understand that his hiring philosophy was never shared. The team was looking for working hands, Adam was looking to expand their family. It was a simple misalignment. Verifying could help ensure that everyone will evaluate the same things.

Adam verified privately afterwards. He explained to the team that he had negative experiences with candidates similar to the one the team rejected. He went back to the culture step in the ACE model and explained, "Caring is one of the core values of our company. The candidate talked very negatively about their previous manager." The team was puzzled. Adam explained, "If you care about someone, you don't complain about them to someone else behind their back. You talk to them directly."

Adam also had a conversation with the team lead explaining that culture is more important to him than technical expertise. If you hire senior experts, they tend to join with experience and learned behaviors that are hard to change. There are core behaviors that Adam wanted to be strict about. Hiring a leader with different behaviors would be an implicit sign that those behaviors were now allowed and rewarded. Also, Adam expected all team leads to understand those core cultural behaviors and make sure everyone on their team followed those values and behaviors.

The team lead challenged Adam. Instead of dismissing the candidate based on what Adam could or could not learn from them, Adam should have told them about his behavioral expectations. Adam accepted it. He was not very explicit about what he wanted from the team lead as he only delegated the task to decide who to hire. He was not very precise on how he thought hiring decisions should be made. Lastly, he was not very coherent on why to reject a candidate, allowing the team to learn how to decide better next time.

Understanding the context for a decision enabled the team to seek candidates who are aligned with the culture Adam wants the company to have. As the CTO, Adam is expected to drive the culture and is expected to care about the culture, so the team appreciated the new knowledge they gained. Furthermore, they no longer felt he has no confidence in them.

Many leaders struggle to explain the importance of certain responsibilities, decisions, and tasks. It can be a result of not knowing themselves or thinking it is implied or common sense. Share your context with others, even when you feel it is obvious. Verify that people understand it the way you meant it. Never assume. Communicate your thoughts and feelings openly and honestly. Challenge your people to tell you when they do not understand you. Challenge them to repeat what they understood.

Remember, repetition is better than not saying anything. If the importance is succinctly expressed, your people will understand the decision-making criteria and principles to optimize their decision on. Also, they will be able to deliver better results.

What Would Be a Successful Outcome?

Next, verifying that everyone understands the expected outcome is vital to ensuring a shared understanding of what success looks like. What do you expect to see as an outcome? What criteria would you use to evaluate how successful it is? What would change when the outcome is successful? When you delegate work to others, they should have the same understanding as you on what would make the outcome successful.

Together with the importance, verifying that people are working towards the same outcome helps prepare their mind for what is happening and what is expected to happen. It helps them prepare mentally for the challenge ahead. This is crucial. People can't read your mind. Make sure you clarify any question they might have. Then verify that you have a shared understanding of what is expected. Doing so will also give you the confidence you are all working toward the same goal.

Let's think again about Adam's hiring challenge. When Adam delegated the hiring responsibility, he gave the team a job specification that included technical and soft skills required for the role. In the past, Adam made all the hiring decisions, so he controlled the culture.

What is Adam's definition of a successful hiring decision? Verifying with the team that a successful candidate should first be aligned with the company culture is vital. It was not obvious to the team. Adam assumed it was a given. The team, however, thought the values were a marketing tool. They could have not known Adam was using them as evaluating tools in every interview he held with candidates before. Also, they could not anticipate that culture is more important than the skillset requirements since the latter were mentioned explicitly in the job specification while culture was never discussed after the onboarding. Now everyone is clear on what successful hire looks like.

How Does the Delegation Look?

How refers to the methodology expected. The answer to *how* questions is often a list of steps that need to be taken. In the context of creating clarity for autonomy, how is a set of guidelines rather than concrete steps to be taken. Here are some questions to think about when you delegate:

- What are your expectations?
- Who should be the final decision maker?
- Are we talking about a full ownership on their side or a partial one?
- Are you there to support?
- Would you like to be involved or informed?
- Should people work together with others?
- Who else should be involved? What information should we gather?
- Would the decision be facts-based or opinion-based?
- How often should you check in?
- How would you expect to measure the success of the final outcome?

These are only a starting point, example questions you could utilize to clarify the how. Considering the type of work, responsibility, or decision you delegated, more questions might come to you or your team as you ask the core question of "How should delegation happen?"

What if you struggle to answer these questions? Then you might want to rethink if you are clear on what you want, why you want it, and why you should delegate it. Also, this step, similar to most steps in this book, can be done together with the person you delegate to. The only question you need to be able to answer yourself is "what are your expectations?" but nothing should stop you from first brainstorming on what they should be. Just don't forget to verify that you reached the same conclusions after you clarified.

When he delegated the hiring decision to the team lead, Adam was not able to answer most of the questions above. Therefore, he invited the team to lead and the team to brainstorm. Together, they came up with a clear process and a shared understanding of how Adam should be involved in the hiring process. More brains are better than one.

Adam also went the extra mile in involving Fabian, the new team lead hire, in the brainstorming. His expertise in how other companies implemented their hiring decision process was highly valuable to shine some light on what collaborating and delegating such a decision could look like.

Before the Verifying step, the hiring decision was delegated as "From now on, the team leads make the final call on who to hire for each role." In practice, it was not the case. Also, this type of delegation was not as explicit, precise, or coherent as it should have been.

Framing it clearly, the hiring decision was delegated as an extension of the team lead role to ensure they found the right person to complement the team. Hence, when hiring a candidate, they first looked for a cultural company fit. Then they evaluated that person's ability to do the job based on the job specification. Also, Adam wanted to be informed rather than continually involved. It wasn't sustainable for Adam to be present in every interview, except for those with very senior candidates. When Adam was involved, his opinion weighed the same as everyone on the team. The final call was made by the team lead. Adam verified everyone understood that.

Case Study Epilogue

Adam understood that creating autonomy is an important fundamental to creating clarity and leading an aligned team. Like any step in this book, his understanding on how to do it evolved over time. He started small by being more transparent and less assuming.

As a leader, you are responsible for creating clarity and ensuring that everyone is aligned. Creating autonomy means that you are not responsible for finding all the answers alone. You are responsible for ensuring that your people can operate independently within the boundaries of the direction you provide. Involving others in each step of the ACE model expresses your trust in the people you work with and the belief that you are smarter together than alone.

At this stage, you understand that clarity is all about asking the questions and bringing people together to answer them. Autonomy is an important stepping stone to ensure that. It is not enough to assume people should know because you feel that you know. Moreover, knowing is different from understanding, and what you want is a shared understanding. Including your people in your thought process is what brings the diverse perspectives to the table and what creates clarity and a shared understanding.

Adam understands that giving autonomy and practicing effective delegation goes together. Now when he walks into the organization's building, he is proud of his people as much as he is proud of everything they have achieved. Autonomy for Adam as a leader means that he can focus on growing the company while having the trust everyone is on board, supporting each other to build a better future than he himself can imagine. Also, it is nice to be able to take vacations.

Conclusion

Autonomy, the ability of one to own their work fully and make decisions about their work, is vital in working environments that are going through continuous and frequent change. Nevertheless, autonomy is not a given. If you want your people to work autonomously, you must invest in both creating the foundation for it and preparing your people for it. You must make sure your people are emancipated and capable of taking over the decision power for areas within the scope and boundaries of their work. Also, ensure that you are ready to delegate effectively. Success in creating autonomy yields rewards in people's motivation and performance, and frees you to work more strategically.

The ACE model begins with Identity, goes through eight major points, and leads up to Clarity. Now that you have gained access to all steps on the ACE model, what is the impact of the new clarity you gained about the way you lead? What is going to change for you going forward? What does leading with clarity mean to you?

Chapter 5 | Autonomy

Questions to get you started:

1. What is in your control?
2. What decisions are part of the areas that are in your control?
3. Who should make the final decision?
4. What makes that person the right one to make that decision?
5. What capabilities are required to make that decision?
6. What is the impact of not delegating this decision?
7. What is the level of confidence you have in that person?
8. What would increase your level of confidence?
9. What makes the delegated responsibility important?
10. Why should you do this task? Why should they do this task?
11. What would be a successful outcome?
12. How should the delegation look like?

Next, we are going to look at how to lead with clarity. Flip to the next chapter and discover more.

CHAPTER 6

Leading with Clarity

Bringing Everything Together to Create Alignment

If you are reading this chapter, it means you have successfully familiarized yourself with all aspects of the ACE model and you are now ready to lead with clarity and discuss what alignment looks like.

When we started, the first concept I introduced in the first chapter was clarity. It is the idea that a large part of the friction we feel in life, at work, and in relationships comes from people seeing the world through their unique lens. Therefore, we also have different understanding, values, needs, and expectations. We make decisions, drive our agenda, and pull our world in a direction that is aligned with that unique lens. When others are aligned with us, we are in harmony. However, when others are not aligned with us, we feel friction. If we are unclear on these differences, we are unclear on what needs to happen to bridge our expectation gap. We struggle to make decisions. We struggle to align. Hence, clarity is a foundation for alignment.

The ACE model manifested to provide a way of thinking and a practical framework for leaders to start asking questions. The ultimate goal is to start the process of clarifying so you can align better. It is about getting into the habit of clarifying rather than assuming you have all the answers. Sometimes you will discover that you know the answers. However, my experience was that there is always much more to explore than I think.

The framework is tailored for digital technology environments, meaning companies that are going through continuous and frequent organizational change due to digital business model transformation, accommodation of digital product development teams, and the growing need for business agility due to market disruption. The type of leadership I advocate for in this book is aligned with the agile mindset and collaborative and shared leadership styles. Nevertheless, I urge you to always read content with a critical mind. Take what is relevant to you, try it out for a couple of months, and judge whether it works for you.

This chapter is a recap. At this point, you explored each of the ACE model concepts as a standalone concept. Most of the time, this would be enough to clarify a situation and align with your people on what to do next. However, to make the best out of this model, I do want to take a different perspective on how the different parts fit together. Ready?

Recapping the ACE Model

The ACE model consists of three main concepts: *autonomy*, *capability* and *emancipation*. As you can see in Figure 6-1, each concept influences one outcome: clarity. Nonetheless, by taking small steps inspired by any one of these three concepts, you are contributing to that outcome. Starting with the end in mind, the last step brings you closest to clarity. With clarity, you can align better.

Leadership in a Time of Continuous Technological Change

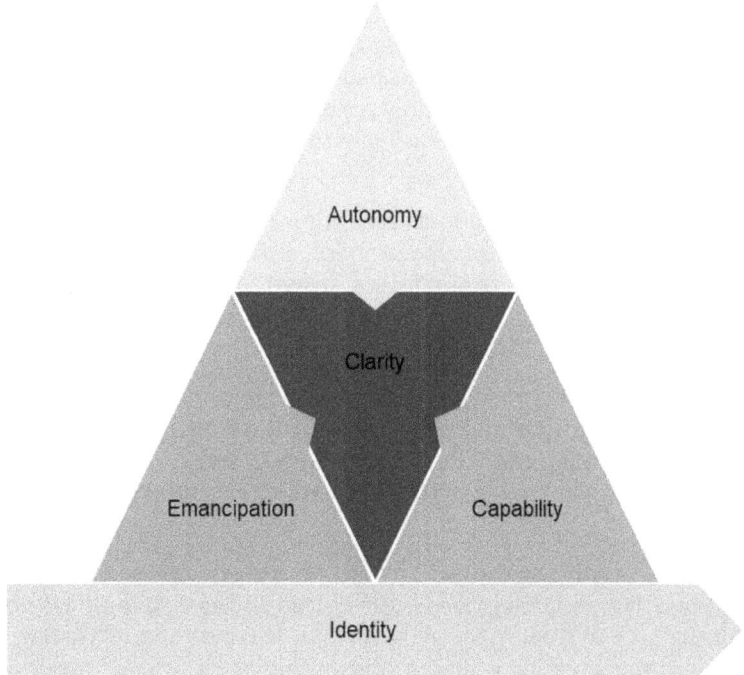

Figure 6-1. The ACE model

When we started, you were introduced to *identity* as the foundation of the ACE model. If you want to change anything, you must start with yourself. You need to fully understand the change that you are driving. Clarity, in this context, starts with you. Every leader should create awareness in themselves before commencing the process with their people. Simply put, you need to know what you want, what you need, and what you need to change to get there. Then you can talk about what others need to change.

I then introduced you to the concept of *emancipation*. My ultimate goal was to encourage you to partner up with your people and emancipate them through the creation of a shared vision and understanding of the purpose, culture, and mission of your organization, ensuring you are creating the space for them to lead and make decisions without dependency on you.

Emancipation means breaking free from having to rely on you. It is about clarifying the boundaries you operate in. It is about making sure that people understand the strategic direction the company is taking, the meaning of what the company is doing, the rules of engagement with other individuals, and prioritizing the most important thing right now. When people are clear on these topics, they can align themselves with the organization and create a sense of meaning in what they do. Also, this is the best way to test the personal fit between the person and the organization.

Next, I explained that building *capabilities* is a joint effort. I introduced you to questions and tools. Through analyzing the structure and process of your organization, you can assess what capabilities people require to succeed in their role. When you use these tools together with your people, you create opportunities for them to gain knowledge and maximize their growth through awareness of their strengths.

Capability, in this context, is about clarifying where someone can contribute their best work. It is also the foundation for most conversations about roles, responsibilities, performance, and future growth. It enables you to share this with the other person so that they can own, over time, their professional development.

Finally, there is the concept of *autonomy*. This concept builds upon the concepts of emancipation and capabilities. It is about delegating control to your people, developing confidence in each other, and ensuring clarity so that everyone understands how everything fits together and what is expected of them.

The main value of utilizing the ACE model to create autonomy is that it helps you to ensure that you delegate in a structured and agreed upon way. It is not about letting go and hoping things work out. It is not about trusting people blindly. It is about creating a shared understanding of who is responsible for what and in what way. It is about clarity. It is about building trust and confidence in others. When your environment keeps changing, clarity on how autonomous people should or can be is vital to ensure the right balance between control and freedom.

The Reality of Transformations

As a leader, dealing with transformations and changes in your workplace can be daunting. I am unsure if I stressed this enough so I will repeat it one last time: changes are now a norm in organizations. This is not a temporary phase; this is an ongoing reality. It is not going to get slower and stabilize. It is going to get faster, but we will have better tools to deal with it.

When changes occur, they can trigger fear and concern in everyone involved. Yes, even if people are early adopters or welcome change. The workplace is one of the areas in life where most people value stability. For example, if there is a modification in a person's role, it might come with feelings of fear of lacking capabilities or losing power, freedom, and control over the situation. Also, many organizations evaluate people based on traditional performance indicators that are defined by their role. Thus, people question how these changes will impact their evaluation, promotion, and future in the company. To be clear, almost every organizational, structural, and process change may require adaption of one's role. In some cases, product direction and strategy change require it. Hence, we constantly balance between the organizational need to change and the employee's need for stability.

The State of Clarity

Clarity is a state where a person knows what they need to do and how they need to do it to contribute to reaching a clear change outcome. Also, they understand and experience no doubt that their contribution is important for the change to succeed or for the outcome to be reached. Otherwise, why do it?

Clarity feels like an "aha!" moment. It is a state of certainty where a person is confident that they understand what they are trying to achieve. They understand how they could or should achieve it and feel empowered and capable to contribute to see it come to fruition. When you lead with clarity, you gift your people and yourself that experience.

Clarity is not about agreement. This is where alignment is important. You can only align when you are both clear. Clarity is the prerequisite for alignment conversations.

The essence of the ACE model is to encourage leaders to proactively engage in the conversations that create clarity on things that matter. The model supports you to ensure that you are not running all by yourself. It ensures that you take people along with you as you lead changes in the organization. That people understand what you want. That you understand what you want. It allows you to be more inclusive and agile because these conversations provide you with important information that you could not get otherwise.

The importance of clarity in leading change cannot be overemphasized.

When you are clear on what you want and need, you can communicate it with others.

When everyone is clear where they are going, they can focus, supporting you and themselves to get there.

When everyone is clear of what is expected of them, they can contribute and bring their best selves to it.

Change Is About People

In trying to build a better future for the organization, you must become aware of the people aspect of every change process. Leading with clarity is first and foremost about recognizing the importance of truly understanding the differences between people. It is about recognizing that change happens through each individual changing themselves. After all, change is a journey. Change is a choice.

People are the most important resource of every organization. When you care for them, they care for you. When you take the time to understand them, they will take the time to understand you. When you are clear with them, they will be clear with you.

Change can only happen through people and with people.

Balancing Control and Freedom

One of the most common misconceptions about agility and ways of working on a digital product development team is that they must operate in chaos. Things are going to change anyway so why bother planning or clarifying things? We figure it out as we go.

What typically happens is that most people struggle to operate under a state of complete freedom. The need for autonomy is often confused as a need for freedom. Autonomy is the feeling that you have a say in how you do your work. That you can make decisions within the boundaries of your role. That you can control the outcomes you are responsible for. It is not to be confused with freedom, the need to do whatever you want.

There is a delicate balance between autonomy (the freedom to decide within your boundaries) and dependency (relying on a boss to decide everything). This balance between autonomy and dependency is highly influenced by the leader's need for control. The more control you need as a leader, the more freedom people seek. The less control you need as a leader, the less freedom people seek.

In a state of complete chaos, leaders with a high need for control will micromanage to compensate for their feeling of uncertainty. It is impossible to micromanage the unmanageable so people will get frustrated. People will struggle to see the bigger picture. Also, you may have to work too much, struggling to finish things and feeling like your people are not taking ownership where they should.

The other side of that spectrum is no better. Leaders with a low need for control are completely hands-off. The most common feeling then is that the leader doesn't care. So no one makes decisions. Such a leader may struggle to understand why people have too much work to do, why certain things are not getting done, and may feel that some people are not doing their work.

The aim of the ACE model is to help leaders shift from firefighting or micromanaging in the chaos of change to explicitly involving people to create the clarity and focus for the change to succeed. I cannot emphasis how often leaders either accept the chaos as normal, while focusing on telling people to follow a master plan that makes no sense to those who have to implement it, or they leave people unsupported completely because they expect them to change things themselves with no real authority to impact anything.

In most cases, leaders are unaware that they do these things. It is the job of leaders to balance their need of control with the need of their people to have the right amount of freedom. This freedom is only possible when you clarify the context and expectations.

Try Things Out

As a leader, when you take the time to observe, you will discover that most of the misalignment and friction that occurs in your organization are either caused by or magnified by a lack of clarity.

Don't just take my word for it. Take five minutes and ask people around you "if nothing changes and we continue on the same path as we are now, where do you see the company, realistically, a few years from now?" Is the answer one you want to hear? How far is it from where you see it?

Do not force it. Try it. Start wherever you are comfortable starting. Take as small a step as you feel comfortable taking. You cannot fail when implementing the ACE model. After all, it is all about asking questions. Worse case, you get to know your people better.

Asking might feel uncomfortable when you are not used to working collaboratively with others. This feeling will fade away as you practice. Nevertheless, you can start with any question so choose the ones you feel comfortable asking.

My rationale behind developing the ACE model as a flow was for you to start with your clarity and end up with others' clarity. Hence, identity is about you, autonomy is about others. Emancipation and capability are often a requirement for autonomy. That is it. Nevertheless, the flow is a recommendation. You can start at every step and finish at any step, depending on where you feel clarity is missing for your people and yourself.

Case Study: My Story

Recall my story back in Chapter 1. I joined a consulting firm after years of working on software development teams. I was hired because they wanted to expand and change the way consulting engagements were conducted. I thought I knew what my role was, until one engagement proved that I did not. The unsuccessful engagement made me realize that I was lacking clarity. You would think that after realizing this, I would take the lesson to heart. I knew what I had to do, but I needed to act on it. In hindsight, I needed a stronger push to do so. Let's use my story to recap the ACE model.

Identity

After leaving the firm, I joined a software engineering team in a new company. Leading software development was my comfort zone. The role, however, was completely new. I was hired to improve the way engineering operated. It built upon my diverse experience, it was aligned with what I wanted to do, and it was a huge step in my career.

It was a conscious choice to leave consulting and move back to software engineering. While I thought working in a large consulting firm would fit me, it was not the case. Short engagements were too short to create the type of change I wanted to create. Also, the hierarchy and leadership style in software companies tend to be more aligned with my preferred leadership style. Starting to clarify my identity contributed to this decision at that time.

Emancipation

At my new company, I repeated the same mistake I made at the consulting firm. Once again, I did not clarify what I was there to do. I jumped right in, assuming I was hired because of my consulting experience. Thus, I thought they wanted a consultant. Consultant is what they got.

I spent my first month in gathering data and getting familiar with every possible tool that included data. I was the expert. I wrote everything down on a slide deck and prepared my master plan. Imagine my surprise when I realized they were not huge fans of long slide decks.

That was when I had a feeling of a déjà vu. I realized that I had an opportunity to make better choices this time. So I went home and thought about it. I wrote down my knowns, unknowns, assumptions, and expectations. The next day, I started asking questions to emancipate myself. I didn't stop until my last day there.

This company was a special case where the leaders could not emancipate me, so I had to emancipate myself. Sometimes you may find that you are not the only one who is unclear. In my case, the leaders in this organization could not answer my questions, so I had to support them in clarifying it for everyone.

There are no shortcuts to becoming an effective leader; the process entails a lot of hard work, consistency, and consideration for the people you work with. I prepared a new plan based on the information I gathered from conversations with people. I talked to everyone possible across all roles and seniority levels. When possible, I backed up my plan with facts, figures, and numbers. Nevertheless, the main challenge was the lack of clarity. Hence, my mission was to bring clarity.

When I presented the plan to change the whole structure of the engineering department and introduce new roles so that we could measure performance better, the reaction was positive beyond expectations. None of the problems I raised were new to my managers. Even the root cause was clear to them. Moreover, they were on board with changing the structure and roles. They were so onboard that they had already thought about it and had great ideas.

For them to improve their ways of working, they needed to have clarity of how things were going to change and have a clear reason for every change. They were mostly rational people. For them, it was not about improving engineering; it was about ensuring everyone could do their work. People were blocked so they either started new work or waited. They were in a constant state of indecision. They had no clarity on the strategy and no alignment between the organizational vision and the day-to-day work. Everyone did the best they could.

We talked about processes all the time. Also, they implemented known agile frameworks and practices. Moreover, they had coaches in place to support them. This was not enough. They needed clarity to fully understand what the change entailed for each individual.

I remembered how things went at the consulting firm. Everything was clear in my head because I made so many assumptions. This time, I decided to make no assumptions. At every step of the way, from clarifying the reasons for the changes to designing the new structure and roles and facilitating the change with the multiple engineering teams, I involved others.

Together with the leadership, we clarified the company values and strategy to link the new structure and roles to the purpose, culture, and mission of the company. We understood that we were missing crucial capabilities to make this happen. Also, it helped that the leadership supported collaboration as one of their primary values.

I learned to accept that collaborative work would take longer. I was so used to driving things myself. I was used to my speed. It was uncomfortable not to decide right away, to take the time to include, involve, and ask questions.

I took it to the other extreme. I asked questions for things I thought were obvious. When they were not, I was surprised. I had my moments of doubt whether we were trying to reinvent the wheel. Often, I felt we discussed the same points repeatedly. At some point, I questioned the importance of the change and their readiness for it. However, when the future structure and roles were clear to everyone involved and we announced them to the whole department, everyone was on board and everyone had a place. We were all emancipated.

Capability

The next step was implementing the change. We shifted people between roles, communicating clearly what was expected of each role and starting with the handover of responsibilities and tasks previously owned by different people to a new person. This step was harder than the previous one because it created a lot of resistance. Wanting the change was one thing, acknowledging the change in practice was a different thing.

Some people realized they had to give up power or responsibilities they did not feel like losing, even if they understood it was the best decision for everyone else. Others were afraid that they were not capable of taking on a new role, even if they understood that they were not expected to know everything on day one. Certain roles kept the same title, so people expected them to remain the same. Also, we did not have all the people we needed in-house and it took time to hire. Thus, some people had to combine a new role with an interim role they never performed before. Once again, everything was normal and expected. Nevertheless, for them, it was the first time, so clarity was even more important. This was when I started leading with clarity.

As mentioned, people were involved in every step of the way, including choosing their role and clarifying it. They also got a say in choosing their new team and process. What the ACE model allowed me to do was have a set of questions to guide the discussions to ensure people provide me with valuable input and developed clarity.

Autonomy

Only few weeks into the implementation, we started noticing the positive impact of the change. Beyond performance, people were happy and took much more ownership than before. One of my managers even said that he delegated so much work, he finally had time to focus on strategic areas.

There was no day zero when everything changed. Some changes happened immediately when the new roles and processes were approved. People transitioned slowly into their new role, supported by training and coaching. Other teams had an interim person supporting them until we hired new people with the required missing capabilities. Over time, the changes became a new reality. It became the new normal of how we did things.

In the beginning, I introduced a time slot where people could ask questions. Over time, it was enough to align with people when they needed it. Also, with the new people on board, my focus changed together with my role. This time, I made sure to keep clarifying and asking what I was there to do.

Leadership in a Time of Continuous Technological Change

Most transformations fail, so I wondered what made this one different. Why did it succeed? I realized that it was because we put people first. The ACE model began to solidify in my head and I knew it had to be shared. And here we are today.

Conclusion

"If you want to go fast, go alone. If you want to go far, go together."

—African proverb

Thank you for taking the time to read the whole book. I hope it has been an informative journey for you. If you wish to provide feedback, ask a question, or need support, please do not hesitate to reach out to me at bar@lead2coach.com.

CHAPTER 7

Leader's Question Guide

The questions that follow are a quick way of internalizing all that you have learned from this book. Whenever you feel like clarity is fading away, check in with yourself by answering these questions.

Identity

1. Who are you?
2. What motivates you?
3. What makes you unique?
4. What do you value?
5. What are your needs?
6. What do you believe to be true above all?
7. How do you see yourself in comparison to others?
8. What type of a person are you at work?
9. What type of a person are you at home?

10. What experiences shaped you to be the unique person you are?
11. What situations trigger your negative feelings?
12. How do you feel about your current leadership position?
13. What does leadership mean to you?
14. What kind of a leader would you like to be?
15. What strengths do you bring to the table?

Emancipation

1. What is your desired impact on the world?
2. What do you need to do to achieve your desired impact on the world?
3. Why do you want to create this impact on the world?
4. What does success look like in your current situation?
5. What would success look like in the future?
6. What behaviors do you value? What behaviors does your organization value?
7. What is considered normal behavior in your organization?
8. What behaviors do you aspire to have but don't necessarily have right now?
9. What is the most important thing right now?
10. What makes it important?
11. What would it look like when you achieved it?
12. How does your contribution or your team contribution support you or your organization in achieving your goals?

Capability

1. What is your current structure?
2. What core processes are you part of?
3. What triggers the process?
4. What is the outcome of the process?

5. What steps are taken to complete the process?
6. Who is involved in each step?
7. What role do they have?
8. What makes this role important?
9. What decisions are owned by this person?
10. What do they need to do to complete the activity or task in this step?
11. What do they need to know to complete the activity or task in this step?
12. What knowledge and skills are required for this role?

Autonomy

1. What is in your control?
2. What decisions are part of the areas that are in your control?
3. Who should make the final decision?
4. What makes that person the right one to make that decision?
5. What capabilities are required to make that decision?
6. What is the impact of not delegating this decision?
7. What is the level of confidence you have in that person?
8. What would increase your level of confidence?
9. What makes the delegated responsibility important?
10. Why should you do this task? Why should they do this task?
11. What would be a successful outcome?
12. What should the delegation look like?

Index

A, B

ACE model, 15, 28
  autonomy, 15
  capability, 2, 15
  clarity, 3, 4
  concept of agility, 11
  consulting engagements, 9
  digital engagement, 10
  emancipation, 2, 15
  entrepreneurial mindset, 1
  fundamental differences, 9
  identity, 16, 18
  impact on work environment
    customer experience, 5
    manual data-gathering processes, 6
    organizational change, 6
    speed of delivery, 6
    transformative process, 7
    use instant messaging software, 7
  keep end in mind, 5
  lacking clarity, 8, 9
  organizational structure, 67
  problem-solve effectively, 1
  shifting mindset, 11–14
  step-by-step instructions, 2
  victim mindset or blaming culture, 10
Autonomy, 106, 107, 111
  ACE model, 15, 95
  Adam's story
    addressing customers' problems, 78
    conducting individual conversations, 78
    connection between control, confidence, and autonomy, 80, 81
    connection between trust and delegation, 79, 80
    coordinating work, 78
    customer expectations, 77
    day-to-day challenges, 78
    growing company, 76
    interviewing potential candidates, 78
    nepotism, 79
    new challenge, 78
    people management, 77
    personal touch, 77
    psychological gap, 79
  bad habits, 76
  confidence, 87–89
  control
    capability, 86
    everyone in control, 82, 83
    giving control, 86, 87
    incorrect performance evaluation, 85
    power to influence priorities, 82
    responsibilities, 84
    understand risks, 85
    you are in control, 83
  creating autonomy, 76
  good habits, 76
  initial start-up phase, 75
  ownership, 76
  practicing effective delegation, 95
  self-governing, 75
  trust, 75
  verifying

Index

Autonomy *(cont.)*
    ACE model, 91
    core cultural behaviors, 92
    delegation, 93
    expected outcome, 92
    hand-holding, 90
    hiring decision, 91, 93, 94
    job interview and training context, 90
    negative experiences, 91
    repetition, 92
    risk of misunderstandings, 89
    use exact language, 90

C

Capability, 100, 106, 110
    building
        knowledge, 68–70
        strengths, 71, 72
        structure, 61
    definition, 55
    graphs and metrics, 73
    high capability, 55
    Jeff and Lea story
        A Corp.'s process and structure, 58
        authority cannot substitute, 60
        capability, 57
        challenge, 59
        field of quality assurance and testing, 58
        lack of engineering capacity, 58
        multiple business and operational metrics, 58
        process cannot substitute, 59
    lack of clarity, 56, 57
    at work, 56
Clarity, 3, 4

D

DevOps, 58, 72
Digital age, 1, 8
Digital technology, 1
Digital technology-led environments, 5
Dunning–Kruger effect, 25

E, F, G, H

Emancipation, 99, 104, 105, 110
    accountability *vs.* responsibility game, 38
    clarifying priorities, 51
    consultants and external support, 36
    contribute to decision-making, 37
    controlling leader, 37
    creates a cause, 44, 45
    creating wealth, 43, 44
    culture
        clarifying culture, 49
        leadership behavior, 48
        virtues influence culture, 47, 48
        written and communicated culture, 46
    desired behaviors, 51
    honest with yourself, 37
    lacking clarity, 46
    learning, 51
    level of clarity, 45
    micromanagement, 36
    mission, 49–51
    Nick's story
        challenge, 39
        hire sales consultant to proffer solutions, 38
        lack of transparency, 41
        new strategy, 39
        product development team, 41
        product manager role, 38
        root cause, 39, 40
        second opinion, 40
    organizational structure, 35
    primary responsibilities, 45
    purpose, 42
    self-esteem, 36
    targeted sales strategy, 46
    traditional hierarchies, 36
    transparent and honest information, 51

I, J, K

Identity, 19, 25, 104, 109
    ACE model, 4, 16, 18
    Ellen's story
        carelessness, 23
        context, 19
        hard-working attitude, 19
        hypothetical scenario, 27
        individual team contributor, 27

Index

lack of awareness, 26
leadership style, 26
meet Sarah, 20
mismanagement, 23
new responsibilities, 20
organizational challenge, 23, 24
ownership and responsibility, 25
planning meetings, 23
root cause of, 24, 25
vs. Sarah, 21, 22
self-awareness, 25
set up realistic trackable, 25
team and work process, 20, 21
training, 26
unfairness, 23
planning conversations, 31
self-discovery, 18
exploration and visualization, 29, 30
mindfulness, 30, 31
personality assessment, 28, 29
self-awareness, 28
self-esteem, 17
viable approach, 18
Information transparency, 8

L

Leading with clarity
ACE model, 98, 99
agreement, 101
"aha!" moment, 101
autonomy, 100, 106, 107
balancing control and freedom, 102
capability, 100, 106
change, 101, 102
digital technology environments, 98
emancipation, 99, 104, 105
identity, 104
transformations and changes in workplace, 100

Leadership

M, N

Mindfulness, 30, 31
Myers-Briggs Type Indicator (MBTI), 28

O, P, Q, R, S, T, U

One-sided transformations, 7
Organizational structure
ACE model, 67
analyzing, 61, 62
clear responsibilities, clear goals, 67, 68
interconnect in workflow/process, 65
product development capabilities, 64, 65
simplified software product delivery process, 62–64
start small and take your time, 66
testing challenge, 67
value of, 66, 67

V

Vautonomy, 2

W, X, Y, Z

Vague mission statements, 43

GPSR Compliance

The European Union's (EU) General Product Safety Regulation (GPSR) is a set of rules that requires consumer products to be safe and our obligations to ensure this.

If you have any concerns about our products, you can contact us on

ProductSafety@springernature.com

In case Publisher is established outside the EU, the EU authorized representative is:

Springer Nature Customer Service Center GmbH
Europaplatz 3
69115 Heidelberg, Germany

www.ingramcontent.com/pod-product-compliance
Lightning Source LLC
LaVergne TN
LVHW010344260326
834688LV00036B/862